Exploring Statistics with the TI-83

Dr. Brendan Kelly

Brendan Kelly Publishing Inc.
2122 Highview Drive
Burlington, Ontario L7R 3X4

Copyright © 1998 by Brendan Kelly Publishing Inc.
All rights reserved. No part of this work may be reproduced or transmitted in any form or by any means, electronic or mechanical, including photocopying and recording, or by any information storage or retrieval system, without permission in writing from the publisher.

ISBN 1-895997-11-9

ATTENTION EDUCATIONAL ORGANIZATIONS

Quantity discounts are available on bulk purchases of this book for educational purposes or fund-raising. For information, please contact:

Brendan Kelly Publishing Inc.
2122 Highview Drive
Burlington, Ontario
L7R 3X4

Telephone: (905) 335-3359 Fax: (905) 335-5104

 Copyright © 1998 by Brendan Kelly Publishing Inc.

INTRODUCTION

WHY IS THE GRAPHING CALCULATOR IMPORTANT?

The widespread acceptance of the graphing calculator as an important tool for the teaching and learning of mathematics is now evolving into a compelling assertion of its necessity. Mathematicians and mathematics educators recognize that this new technology, though not a panacea, is an essential mathematical tool that is changing irrevocably the content and pedagogy of high school and college mathematics.

WHY IS THIS BOOK NEEDED?

It is not feasible to expect students to master all the modes and menus of the TI-83 before applying it to the study of statistics. On the contrary, it is important that instruction in statistics draw upon the calculator functions as needed, neglecting those aspects that are not fundamental to the conceptual development. Achieving this integration of machine and mathematics requires a sequence of carefully designed instructional activities that showcase a technology-conscious approach to each statistical concept. This book is designed to be such a resource.

TO THE STUDENT

Your TI-83 calculator and this book will provide an exciting opportunity to discover many of the important concepts in statistics. You will investigate trends in the ages of academy award-winning actors and actresses. You will analyze whether women will outrun men in the Olympics of the 21st Century and you will simulate coin tossing experiments and determine whether the dice are loaded. We encourage you to invest time completing the exercises and investigations, checking your solutions against those in the back of the book and reflecting upon the concepts you encounter. When you have completed the explorations in this way, you will have in your grasp most of the important ideas in mathematical statistics. In your journey toward that destination, we hope you enjoy the content, the cartoons, and the entire experience!

TO THE TEACHER

Drawing correct inferences from the analysis of data is the cornerstone of the new literacy of the information age. Helping students acquire this skill is a daunting challenge to any teacher, because statistical concepts are sophisticated. With technology like the TI-83, the temptation to abandon the quest for deep understanding and to merely train students to operate the calculator as a "black box" is ever present. In this book, we present a careful development of the underlying concepts and statistical techniques *before* we show how the tests and distributions can be accessed on the TI-83. In this way, we hope to help you engage your students in the deep and challenging ideas that underpin this powerful branch of mathematics. We hope you find this book a helpful resource in achieving this goal.

Copyright © 1998 by Brendan Kelly Publishing Inc.

Table Of Contents

Unit 1 One Variable Statistics

Unit 2 Two Variable Statistics

Unit 3 Experimental Probability & Simulations

Table Of Contents

Unit 4 Distributions and Their Properties

Unit 5 Hypothesis Testing & Confidence Intervals

This book attempts to teach mathematics and the use of the TI-83 Graphics calculator at the same time. To do this effectively, it has been necessary to simulate the calculator keys and fonts on the TI-83 and to generate screen displays. Achieving these elements required early access to beta versions of the TI-83 and the TI-83 Link. We are indebted to Len Catleugh, Jeff Crump, and Tom Ferrio of *Texas Instruments* whose continuing help and support in providing early versions of software and hardware have made these publications possible.

Addison-Wesley Canada under the leadership of Tony Vander Woude has generously lent their support to this project. The knowledge and expertise of Anthony Leung, Melanie van Ransberg, Linda Allison, and Roberta Dick have significantly enhanced the esthetics and functionality of this final product.

The covers of the three books in this trilogy are the creation of Gord Pronk of *Pronk&Associates*. They are testimonials to his unique creative talent.

We acknowledge our debt to David Bernklau of Stuyvesant High School in New York City for his painstaking editing of the manuscript. Without his help, this book would have contained many omissions, glitches and goops. As in our other books, we have hidden a few more for you to find, but we trust they have been cleverly disguised. A special thanks to my wife, Teri, who purged the manuscript of a myriad typos, inconsistencies and non-sequiturs.

For a resource to be truly effective in the classroom, it must be free of ambiguities. Instruction must be clear and activities must be sequenced so that students can learn easily from it. The only way to achieve this is to conduct extensive field testing. We are indebted to the students enrolled during the 1993-94, 1995-96 and 1996-97 academic years at OISE/UT (University of Toronto), for working through versions of earlier drafts of the manuscript and providing innumerable improvements. In particular, we express our gratitude to former student Ian Bain who labored through most of the exercises and contributed several useful suggestions for clarifying the presentation.

UNIT 1 ONE VARIABLE STATISTICS

THE STEPS IN ANALYSING & PLOTTING ONE-VARIABLE DATA

STEP 1 ENTER YOUR DATA INTO A LIST

To enter your data into a list, press: **STAT**

We get this display… then press **ENTER** to obtain

```
EDIT CALC TESTS
1:Edit…
2:SortA(
3:SortD(
4:ClrList
5:SetUpEditor
```

```
L1      L2      L3     1
------  ------  ------

L1(1)=
```

Then enter your data into a list in the STAT matrix.

IF YOU WANT ONLY A DATA PLOT, SKIP TO STEP 3.

STEP 2 OBTAIN ONE VARIABLE STATISTICS

To obtain statistics on one variable data, press:

STAT **▶**

We obtain this display… and then press **ENTER** to obtain:

```
EDIT CALC TESTS
1:1-Var Stats
2:2-Var Stats
3:Med-Med
4:LinReg(ax+b)
5:QuadReg
6:CubicReg
7↓QuartReg
```

```
1-Var Stats
```

Enter the appropriate listname (e.g. L_2) and the frequency list (e.g. L_3).

```
1-Var Stats L2,L
3
```

You get a display like this.

```
              1-Var Stats
mean →         x̄=479.4280413
               Σx=235121100
               Σx²=1.20535E11
sample standard deviation → Sx=126.208828
               σx=126.2086994
               n=490420
               minX=200
first quartile → Q1=400
median →         Med=500
third quartile → Q3=550
               maxX=750
```

STEP 3 DEFINE A DATA PLOT.

To obtain a data plot, press:

2nd [STAT PLOT]

You obtain a display like this.

```
STAT PLOTS
1:Plot1…Off
   L1   L2
2:Plot2…Off
   L1   L2
3:Plot3…Off
   L5   L6
4↓PlotsOff
```

Highlight the plot you wish to define and press **ENTER**.
Select from the menu the option **On** and the type of graph.
Enter the names of the **Xlist** and **Frequency list.**

```
Plot1 Plot2 Plot3
On Off
Type:
Xlist:L2
Freq:L3
```

STEP 4 DISPLAY THE DATA PLOT.

To set an appropriate window, press:

WINDOW

Then set the window variables.

```
WINDOW
 Xmin=200
 Xmax=800
 Xscl=50
 Ymin=0
 Ymax=70000
 Yscl=1
 Xres=1
```

To display the data plot, press **GRAPH**.

If the data plot does not appear or fill the display window, press:

ZOOM **9**

Copyright © 1998 by Brendan Kelly Publishing Inc.

Males OutPerform Females on Mathematics SAT

1990 PROFILE OF MATH SAT SCORES FOR MALES

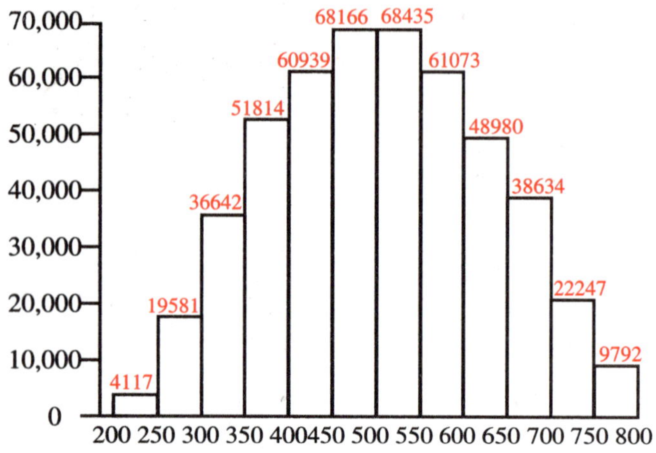

1990 PROFILE OF MATH SAT SCORES FOR FEMALES

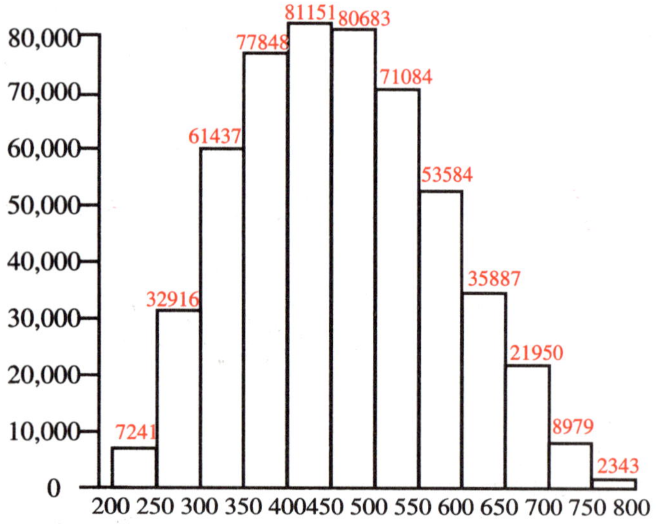

Entrance to American Colleges and Universities is determined by performance on such tests as the SAT (Scholastic Achievement Tests). The following table shows the number of males and females in 1990 whose scores on the math SAT fell within each of the given intervals. For example, the first row of the table shows that there were 4117 males and 7241 females who scored between 200 and 250 (including 200, but not including 250) on the Math SAT in 1990.

Interval	Number of Students	
	Male	Female
$200 \leq x < 250$	4117	7241
$250 \leq x < 300$	19581	32916
$300 \leq x < 350$	36642	61437
$350 \leq x < 400$	51814	77848
$400 \leq x < 450$	60939	81151
$450 \leq x < 500$	68166	80683
$500 \leq x < 550$	68435	71084
$550 \leq x < 600$	61073	53584
$600 \leq x < 650$	48980	35887
$650 \leq x < 700$	38634	21950
$700 \leq x < 750$	22247	8979
$750 \leq x < 800$	9792	2343
Total	**490,420**	**535,103**

Source: College Board 1991

The information in the table above is displayed for males and for females in the *histograms* on the left. The *histogram* above is a graph that shows the number of males in 1990 whose scores on the Math SAT fell within each interval given in the table. For example, we see that the vertical bar representing SAT scores in the interval given by $400 \leq x < 450$ has a height of 60939, indicating that there were 60,939 male students with scores in this interval. The corresponding histogram for female students is shown opposite.

When we compare the two histograms, we observe that the male scores seem to be distributed symmetrically about the line corresponding to $x = 500$. The female scores appear to be shifted to the left slightly, indicating a greater proportion of females below the score of 500. In interpreting these results, we must be careful not to reach quick conclusions about the relative aptitudes of males and females in mathematics. Some groups have charged that the SAT tests are *biased* against females. In 1989, a New York court disallowed the awarding of scholarships based entirely on SAT scores. The College Board has subsequently revised the SAT in attempts to minimize gender or racial bias.

Copyright © 1998 by Brendan Kelly Publishing Inc.

WORKED EXAMPLE 1

a) Construct separately and together the histograms shown on page 8.

b) Construct the histogram of the SAT scores for males using class interval size 100. Trace along the histogram to determine the number of males who scored between 500 and 600 (including 500 but not 600).

SOLUTION

a) The **first step** in constructing any data graph on the TI-83 is to form a table of values. To access a blank table, press: **STAT** **ENTER** .

The x values in the table will be the SAT scores at the midpoint of each class interval. That is, for the class interval $200 \le x \le 250$ we enter the value 225. We define the set of all such midpoints as the list,

$$L_1 = \{225, 275, 325, 375, 425, 475, 525, 575, 625, 675, 725, 775\}$$

To enter these values in the L_1 column of the table, we position the cursor below L_1 using the arrow keys. We then type 225 and press: **ENTER**. We continue in this way until all the elements of L_1 have been entered.

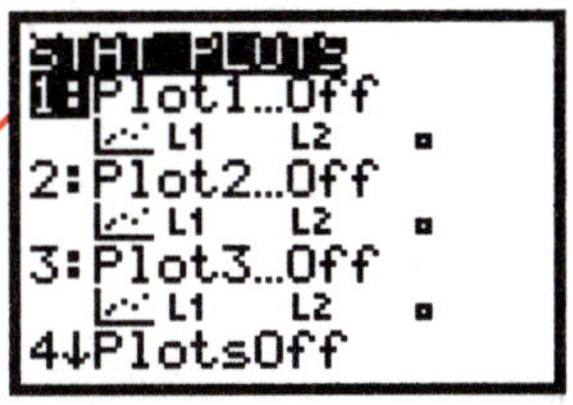

Proceeding as above, we enter the elements of L_2 and L_3, the frequencies of SAT scores in each interval for males and for females respectively. (These are taken from the table on page 8.)

The **second step** in constructing a data graph is to define the statistical plot we want. To do this, we press: **2nd** [STAT PLOT]. We obtain a display like the one shown here.

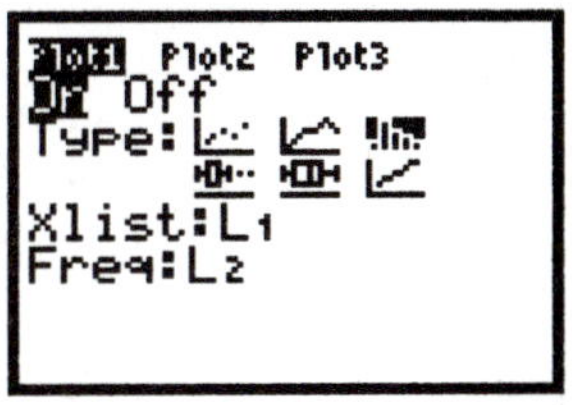

To define Plot 1 as the histogram showing the SAT scores for males, press: **ENTER** and select the options that are shown highlighted in the display on the right. Observe that the histogram is the rightmost icon in the line titled "Type", and L_1 is the x-list. Press **2nd** [L_2] to enter L_2 as the list of frequencies.

The **third step** in constructing a data graph is to define the viewing window. Press: **WINDOW** and then set the values as shown in the display on the right. Then we press: **GRAPH** to obtain the histogram for SAT scores of males, below left.

To create the corresponding histogram for SAT scores of females, we proceed as above defining Plot 2 so that L_1 is the x-list and L_3 is the frequency list. Set Plot 1 to "Off" and press: **GRAPH** to obtain the histogram shown in the middle. Set Plots 1 and 2 to "On" and we obtain the superimposed histograms shown below right.

Note: Xscl defines the size of the class intervals. Its value must be chosen to yield fewer than 48 intervals. Failure to set **Xscl** appropriately will result in an error message!

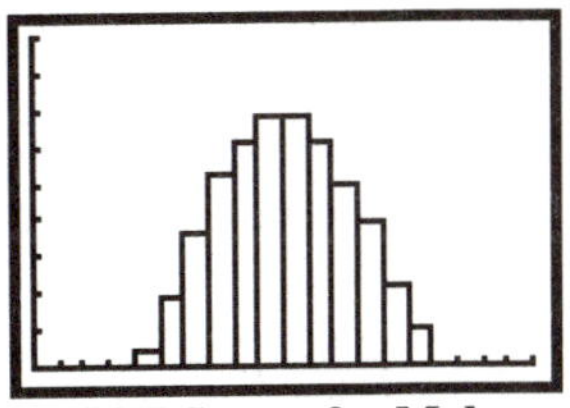

SAT Scores for Males

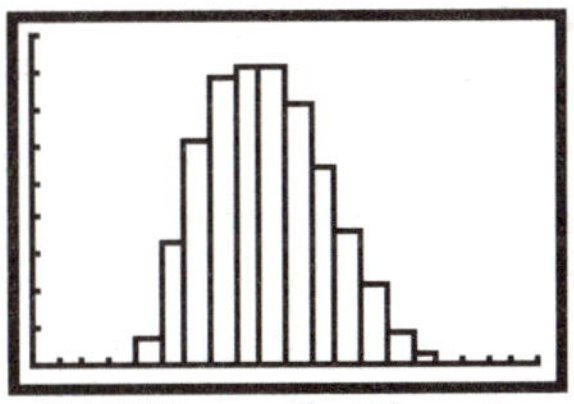

SAT Scores for Females

Superimposed Histograms

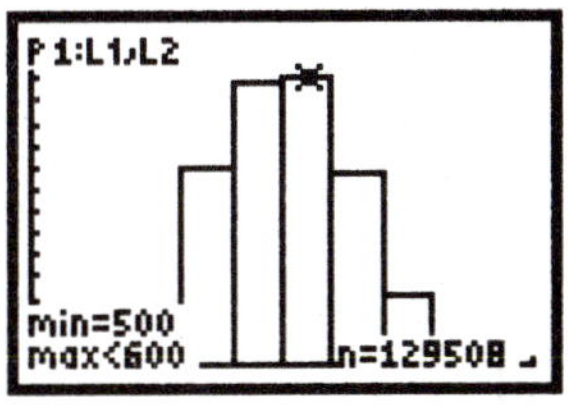

b) To obtain the histogram for the Math SAT scores for males using a class interval size of 100, we merely press: **WINDOW** and change **Xscl** to 100 and **Ymax** to 150000. We then press **GRAPH**. To determine the number of scores in the range $500 \le x < 600$, we press: **TRACE** and move the cursor to this interval. We obtain the display on the right. From this display, we see that there were 129 508 scores between 500 and 600.

Copyright © 1998 by Brendan Kelly Publishing Inc.

WORKED EXAMPLE 2

a) Create a list, L_4, which gives the *cumulative* frequencies of the SAT scores for males. That is, $L_4(n)$ is the total frequency of all scores in the first n class intervals where the nth class interval is defined to be the interval: $200 + 50(n - 1) \leq x < 200 + 50n \ : n > 0$.

b) Construct an *xy*Line showing the percentage of all males whose scores were in the first *n* class intervals. Trace along your xyLine to find the percentage of males whose Math SAT scores were less than 600.

SOLUTION

a) We redefine L_1 to replace the midpoints of the class intervals with their upper bounds:
L_1 = {250, 300, 350, 400, 450, 500, 550, 600, 650, 700, 750, 800}

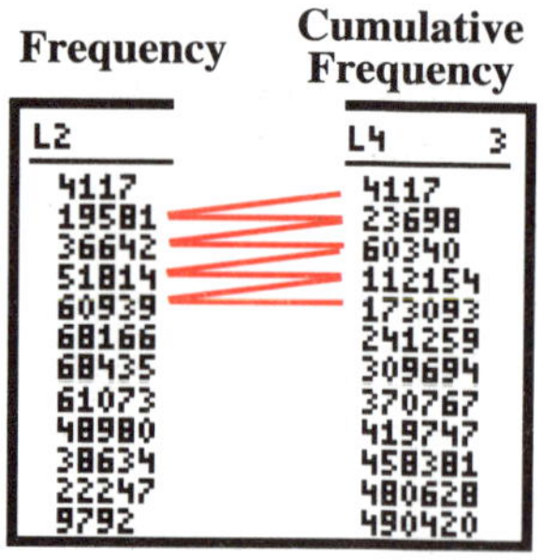

The corresponding frequencies of scores (for males) in these intervals are given by L_2. The *cumulative* frequencies are a running total, L_4, of the numbers in list L_2. To compute the elements of L_4, we use the cumulative sum command, **cumSum(** on the list menu. We press:

2nd [LIST] ▶ 6 2nd [L_2]) STO▶ 2nd [L_4] ENTER

They keying sequence above yields the display shown here.

```
cumSum(L2)→L4
{4117 23698 603...
```

Upon pressing: STAT ENTER we obtain the table showing L_4.

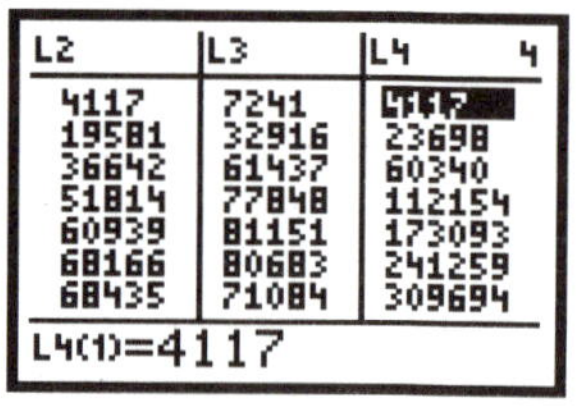

b) To create a list, L_5 such that $L_5(n)$ is the *percentage* of marks in the first *n* class intervals, we define $L_5(n) = 100L_4(n)/490420$. To do this, we enter $100L_4(n)/490420$ on the home screen and store this expression to L_5 as shown in the display.

```
100L4/490420→L5
{.8394845235 4....
```

Alternatively we can define L_5 by returning to the STAT EDIT screen and placing the cursor at the top of the fifth column so that L5 is highlighted, and pressing:

1 0 0 × 2nd [L_4] ÷ 490420

To plot the *xy*Line with L_1 as the **Xlist** and L_5 as the **Ylist**, press:

2nd [STAT PLOT] 4 ENTER This turns off all the statistical plots. Then press: 2nd [STAT PLOT]

We select Plot 3 and then we choose the second icon in the "Type" row (to plot the *xy*Line) and select all other settings as shown in the display below left. To set the window variables so that the entire xyLine fits in the window, and to plot the *xy*Line, press: ZOOM 9.

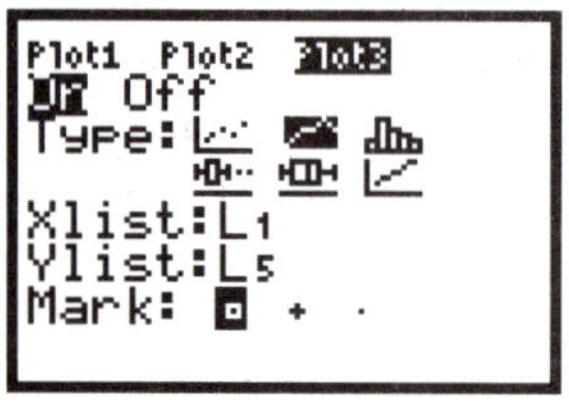

This yields the graph shown in the display on the right. When we press: TRACE and move the cursor along the curve to $x = 600$, the display shows **Y** = 75.60...indicating that about 75.6% of all the SAT math scores by males were less than 600. Such a cumulative percentage curve is called an *ogive curve*.

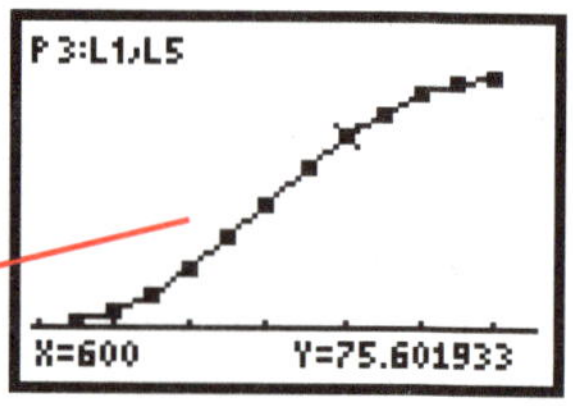

Copyright © 1998 by Brendan Kelly Publishing Inc.

Exercises

1. Use the histogram on page 8 showing the profile of math SAT scores for males to answer the following questions.

a) How many males scored below 250 on the SAT?

b) How many males scored below 400?

c) What percent of male respondents scored below 400?

d) What percent of males scored 700 or above?

e) Can you determine from the histogram or table what percent of the male respondents scored above 700? Explain your answer.

2. a) Follow the procedures in *Worked Example* 1 to re-create the table shown in that example. Check that the numbers in columns L_1, L_2, and L_3 of your table agree with the numbers in the given table.

b) Define L_4 so that $L_4 = L_2 + L_3$. Use your table from part a), and the procedures outlined in *Worked Example 1*, to create a histogram showing the total frequencies of the SAT scores of males and females combined. Use the same intervals as in the table. (These are called the *class intervals*.)

c) Change your plot from a histogram to an *xy*Line by selecting that option from the appropriate plot on the STAT PLOTS screen. This graph is called a *frequency polygon.*

d) Change the histogram so the class interval size is 100; i.e. $200 \le x < 300$, $300 \le x < 400, \ldots$ $700 \le x < 800$. What proportion of the students had SAT scores in the 500s?

e) What proportion of the students scored less than 600?

3. The table shows the number of golfers on the PGA Tour in 1997 whose total winnings (as of September 15) were in each \$100,000-dollar range between \$200,000 and \$1,500,000. (Tiger Woods was at the top with winnings of \$1,949,920.)

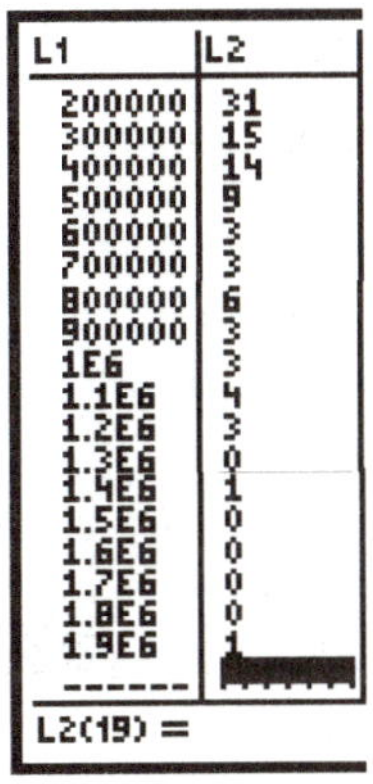

L1	L2
200000	31
300000	15
400000	14
500000	9
600000	3
700000	3
800000	6
900000	3
1E6	3
1.1E6	4
1.2E6	3
1.3E6	0
1.4E6	1
1.5E6	0
1.6E6	0
1.7E6	0
1.8E6	0
1.9E6	1

L2(19) =

L_1 is the list of lower limits (in dollars) of the class intervals. L_2 is the number of golfers in the interval with lower limit given in L_1. For example, the first row of the table shows that there were 31 golfers whose 1997 money winnings (as of September 15, 1997) were in the range:

$$\$200,000 \le x < \$300,000.$$

a) Enter these data into a table like this. Display these data in a histogram. Construct the corresponding frequency polygon. (See exercise 2c.)

b) What percentage of the \$200,000 (or over) money winners on the PGA Tour won a half a million dollars or more?

c) Change the histogram so the class interval size is 200 000. What percentage of the golfers in the top 96 money-winners had 1997 winnings in the interval $\$500,000 \le x < \$700,000$?

Investigations

4. Define list L_1 in your STAT table as in *Worked Example* 2. Define L_2, L_3, and L_4 as in Exercise 2. Then define L_5 so that it gives the cumulative frequencies of the SAT scores for all students. (Follow the procedures in *Worked Example* 2.)

a) Define L_6 so that it gives the cumulative percentages of students scoring below any upper bound. Construct the cumulative percentage curve (ogive curve) for the SAT math scores of all students.

b) Use your ogive curve to determine the percent of all students whose Math SAT scores were less than 600. Compare this answer with your answer to part e) of Exercise 2.

c) In what interval is the ogive curve steepest? Explain why.

5. Construct the ogive curve for the money winnings of the top golfers as given in the table of Exercise 3. Use a class interval size of 100 000. Remember to define L_1 as the list of *upper* bounds in the class intervals. That is, the first class interval $200,000 \le x < 300,000$ corresponds to 300,000 in L_1.

a) Trace along the ogive curve to determine what percent of golfers in the "\$200,000-plus" category won more than half a million dollars in 1997. Compare your answer to your answer to part b) of Exercise 3.

b) Describe the shape of the ogive curve. Explain why it flattens out for large values of *x*.

Note: When creating the cumulative frequencies L_3, remember to select **sequence** mode and define u(*n*) as shown in *Worked Example 2*. Also set u(*n*Min) and set *n* to run from 1 to 13 (number of class intervals). Define L_3 using the assignment seq(u(*n*), *n*, 1, 13, 1) → L_3.

DRAWING INFERENCES

A prestigious university selects only those students whose Math SAT scores are in the top 10% of all those who write this test. Estimate what score a student would have needed in 1990 to satisfy this university's entrance requirement.

In view of the differential between the performance of males and females on this test, could the selection criterion described above be considered sexist (i.e. biased against females)? Explain your answer.

EXPLORATION 2 Central Tendency & Box Plots

The organization chart for High Tech Widget Inc. is presented below. It shows the number of employees in each position and the corresponding annual salary. During salary negotiations, management asserts that the average salary of the workers is \$45,673. The Union asserts that the average salary is only \$36,600. Who is telling the truth?

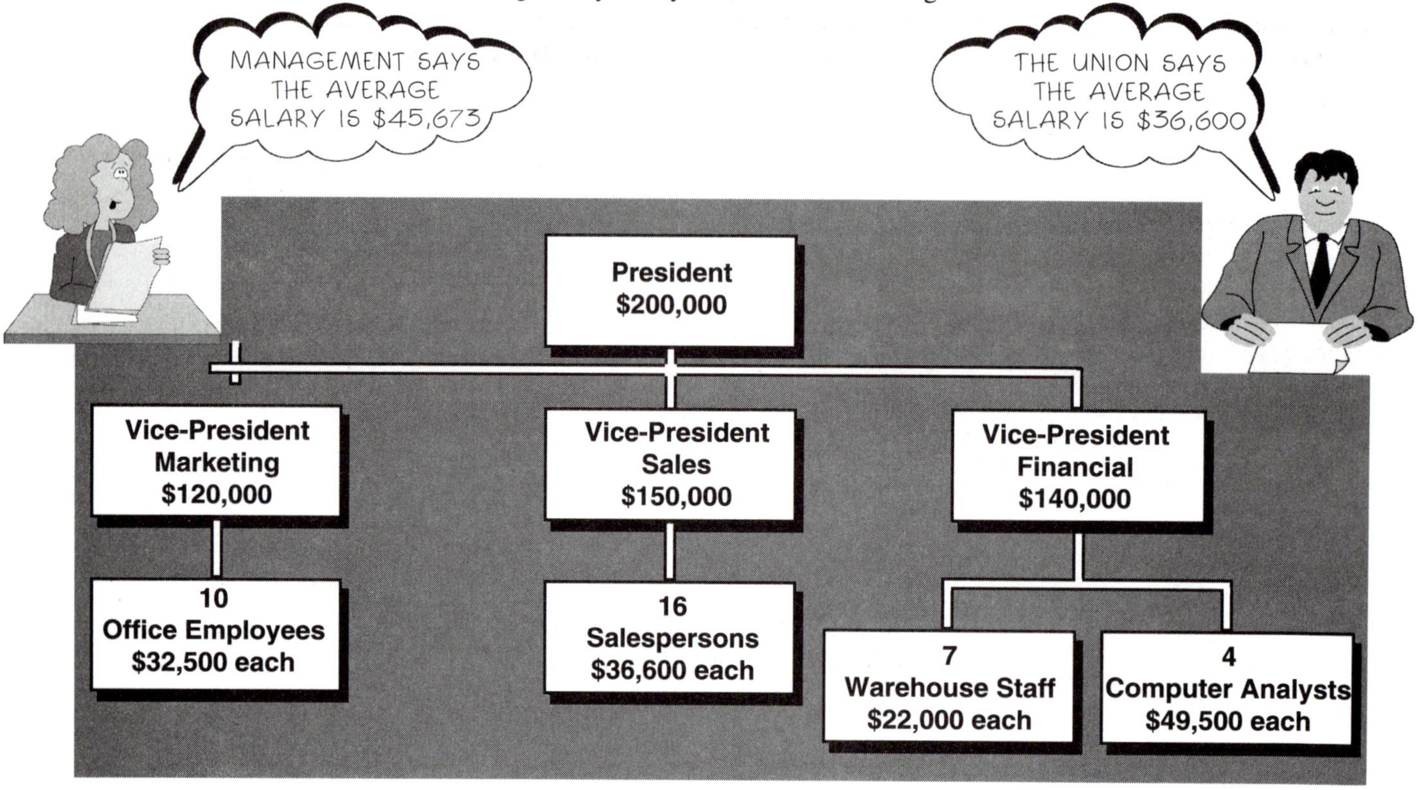

Upon closer investigation, it was discovered that the two sides were using different definitions of average. Management used the *mean*, which is the total money that the company pays in salary, divided by the number of employees. This is the amount of money which each employee would earn if all employees were paid equally. In general, the *mean* (denoted $\bar{x}$) of a set of n numbers, $\{x_1, x_2, x_3, \ldots x_n\}$ is:

$$\bar{x} = \frac{x_1 + x_2 + x_3 + \ldots + x_n}{n}$$

The Union was using the *median* salary as their measure of the average. If the salaries of the 41 employees at High Tech Widget Inc. are arranged in descending order from the president to the lowest paid worker, the *median* salary is the 21st salary in that list. There are the same number of workers earning as much or more than this salary as there are earning as much or less than this salary. In general, the median of a set of n numbers $\{x_1, x_2, x_3, \ldots x_n\}$ is the middle number when they are arranged in ascending order. If n is even, there are two "middle numbers" and the median is taken to be half their sum.

The median divides a set of data into a "lower half" and an "upper half". The median of the lower half is called the *first quartile* (denoted Q_1) and the median of the upper half is called the *third quartile* (denoted Q_3). In the data set $\{x_1, x_2, x_3, \ldots x_{11}\}$, where $x_1 < x_2 < x_3 < \ldots < x_{11}$, the quartiles, Q_1, Q_2 and Q_3 are respectively x_3, x_6 and x_9.

$$x_1 \quad x_2 \quad \underset{Q_1}{x_3} \quad x_4 \quad x_5 \quad \underset{Q_2}{x_6} \quad x_7 \quad x_8 \quad \underset{Q_3}{x_9} \quad x_{10} \quad x_{11}$$

In the set of 41 salaries given above, the first quartile is the salary below which one quarter of the salaries lie. That is, if the 41 salaries were arranged in ascending order, the mean of the 10th and 11th salaries would be the first quartile. Similarly, the mean of the 30th and 31st salaries would be the third quartile. In general, one quarter of all the values in a set of data are at or below the value of the first quartile, one half of the values are at or below the median, and three-quarters are at or below the third quartile.

Copyright © 1998 by Brendan Kelly Publishing Inc.

WORKED EXAMPLE

a) Display the salaries of High Tech Widget Inc. in a table showing the number of people earning each salary amount.

b) Find the mean salary for High Tech Widget Inc.

c) Find the median salary. Define the lower and upper quartiles and calculate these quartiles for the list of salaries. Display the lower and upper quartiles and the median salary on a box plot.

SOLUTION

To clear lists L_1, L_2, L_3 from memory, enter the command **ClrList** L_1, L_2, L_3

a) The **first step** in calculating any statistics is to enter the data in a table. To do this, we press: STAT ENTER .

We then enter the salaries in L_1 and their corresponding frequencies in L_2 to obtain the table shown on the right.

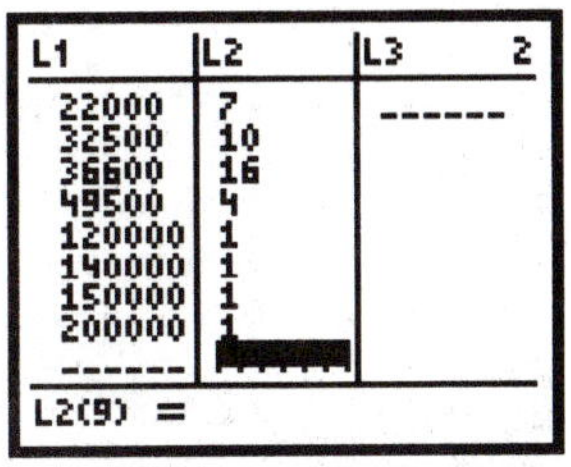

b) The **second step** in calculating a one-variable statistic is to access the **1-Var Stats** command by pressing: STAT ▶ ENTER
We then enter the data list L_1 followed by a comma and the frequency list L_2 as shown here.

Upon pressing ENTER we obtain the display shown on the right.

We see that the mean $\bar{x}$ is equal to $\bar{x}$ = 45,673.17. That is, the mean salary is about $46,673 as stated by the management.

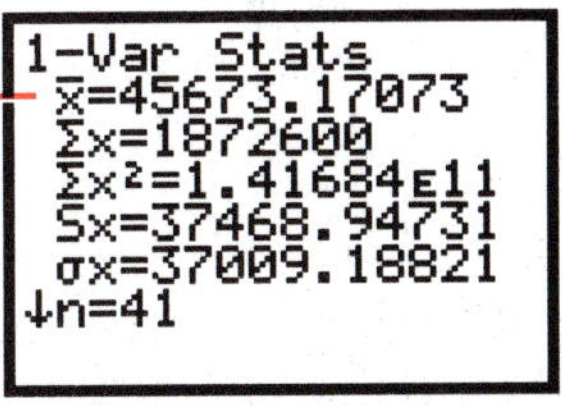

c) The arrow on the display shows that we can scroll downward to obtain more data. This display shows that there are 41 salaries ($n = 41$) and that the salaries range from $22,000 (**minX=22000**) to $200,000 (**maxX=200000**).
The median salary is $36,600 (**Med=36600**) as stated by the Union.

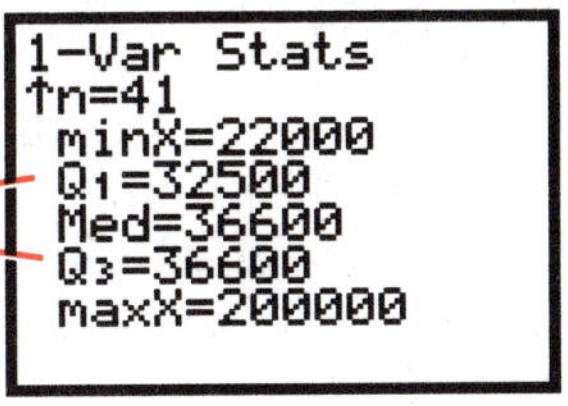

The symbols $\mathbf{Q_1}$ and $\mathbf{Q_3}$ represent the *first quartile* and *third quartile* respectively.
The first quartile is the salary below which one quarter of the salaries lie. That is, if the 41 salaries were arranged in ascending order, the mean of the 10th and 11th salaries would be the $32,500. Similarly, the mean of the 30th and 31st salaries would be $36,600. The median salary is also $36,600, as stated by the Union. (Why are the median and the third quartile equal?)

To create a box plot, press: 2nd [STAT PLOT] 4 ENTER This turns off all the data plots.
To access plot1 press: 2nd [STAT PLOT] ENTER. Then select the options shown in this display.
Upon pressing ZOOM 9 we obtain the box graph shown in the display below left.
By tracing along this graph we can verify the values of the quartiles and the extremes. If we had selected the first box plot icon opposite "Type", we would have obtained the modified box plot (ModBoxPlot), which plots all outliers (data points that are distant from the box by more than 1.5 times the width of the box). See the display below right.

$0 \le x \le 250\ 000$
Xscl = 25000

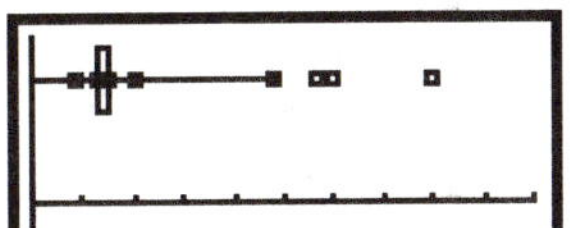

$0 \le x \le 250\ 000$
Xscl = 25000

Copyright © 1998 by Brendan Kelly Publishing Inc.

Exercises

1. Why are measures of central tendency such as the mean and the median used?

2. Suppose a single position paying $180,000 were created at High Tech Widget Inc. Which average salary, the mean or the median, would be increased more by the inclusion of this new salary?

3. Use your calculator to find the mean, median, first quartile, and third quartile of each of the following sets of numbers. (To clear a previous list L_n execute the command **ClrList** L_n.)

a) {15, 16, 29, 36, 45, 16, 18, 29, 16, 29, 32, 36}
b) {-17, 61, 42, 156, -88, -61, 42, 42, 61, -17, -17}

4. *How many dimples on a golf ball?* Most people do not realize (and some do not care) that a golf ball without dimples would not travel very far. The dimples give it aerodynamic properties that increase the distance it travels. Engineers experiment to determine the size and number of dimples that yield optimum distance. A recent issue of a popular golf magazine published the number of dimples on 32 of the most popular brands of golf ball (shown in the table below).

Number of Dimples on each Golf Ball

440	432	392	432	318	442	432	432
432	360	360	360	432	492	440	392
392	492	332	422	422	422	332	440
392	332	432	432	392	332	392	360

Find the mean and median number of dimples on the 32 golf balls given in the table above…

a) by entering each of the 32 numbers into list L_1 in the STAT editor and using the 1-Var Stats as in the *Worked Example*.
b) by making a tally like this and using the procedure in the worked example.

318	332	360	392	422	432	440	442	492
/	////	etc. …						

Check that n = 32 on the **1-Var Stats** display to ensure that you have entered all the data points.
c) Construct a box plot to show the quartiles and extremes.

5. During the 1970's the mean income in a southern state increased by 115% for non-whites and 72% for whites. About 80% of the people in this work force were white. What was the mean increase in income for both groups combined?

6. Determine the mean and median of each set of numbers.
a) {1, 2, 3, 4, …99} (the natural numbers from 1 to 99)
b) {1, 2, 3, 4, …100} (the natural numbers from 1 to 100)
c) {-8, -7, -6, -5, -4, -3, -2, -1, 0, 1, 2, 3, 4, 5}

Investigations

7. The table of SAT scores on page 8 gives the number of students who achieved scores in various intervals. Since the *exact* scores of individual students are not known, we cannot compute the median score from this information. However, we can approximate the median score if we assume each student received the middle score in each interval. That is, we assume the 4117 male students and 7241 female students who achieved scores in the interval $200 \le x \le 250$ all achieved scores of 225.
Making this assumption, construct box plots showing the quartile and extreme scores for the males and for the females on the SAT. Compare the shapes of the two box plots.

8. In a study of the longevity of a particular species of cat, biologists recorded the lifespans of 30 cats. Their results are presented in the following table.

Lifespans of Cats (in years)

12.9	13.2	14.1	13.9	12.8	13.1	13.1	13.2	13.6	13.0
13.4	13.6	12.9	13.3	11.8	12.8	14.6	12.8	10.4	14.8
11.5	13.5	13.6	12.9	9.6	14.5	13.5	13.8	14.4	13.3

In order to enter these data into a list, it is useful to display them in a *stem-and-leaf diagram* in which the stem is the whole number and the leaf is the decimal digit.

Stem	Leaf	
9	(6)	This stands for 9.6
10	4	
11	5 (8)	This stands for 11.8
12	8 8 8 9 9 9	
13	0 1 1 2 2 3 3 4 5 5 6 6 6 8 9	
14	1 4 5 6 8	

Use the information in the stem-and-leaf plot to find the mean and median lifespans in years. Create a box plot and trace to find the first and third quartiles.

CHALLENGE

A doctor lists all the diseases that afflict one or more of his patients. Opposite each disease he records its frequency among his patients. That is, he records the number of patients that have each of these diseases.

The set of frequencies for all the diseases on his list has an arithmetic mean of 20. If the flu, which has a frequency of 92, is removed from the list, the arithmetic mean of the remaining frequencies is 18.

How many diseases remain on the doctor's list ? What is the largest possible frequency for any disease that remains on the list?

Copyright © 1998 by Brendan Kelly Publishing Inc.

Do Great Actresses Mature Earlier Than Great Actors?

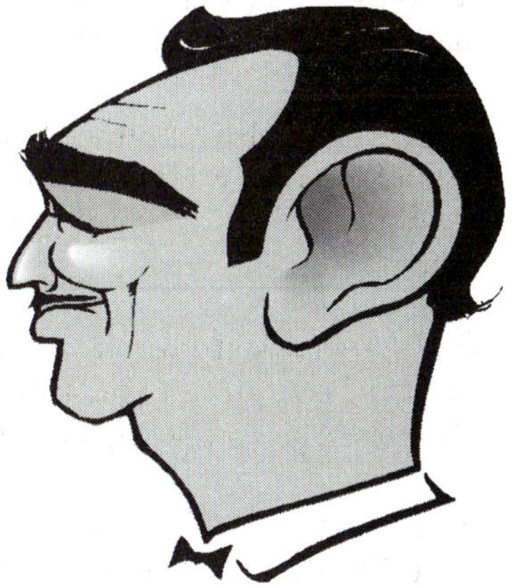

Questions like the one above are difficult to answer because the answer usually depends upon the individuals who are compared. Such questions are often rephrased as "Does the average actress reach maturity faster than the average actor?". Again the question is meaningless because it assumes the existence of a fictional entity called an "average actress". The following question is better defined and more measurable, though it does not address the original question: "How does the average age of the Oscar winners for best actor compare with the average age of the Oscar winners for best actress?"

The tables on pages 15 and 16 show the ages of the winners of the "best actress" and "best actor" awards in the 50 years from 1942 through 1991.

"BEST ACTRESSES" DURING A HALF CENTURY— 1942-1991

Year	Actress	Age	Movie
1942	Greer Garson	34	Mrs. Minever
1943	Jennifer Jones	24	The Song of Bernadette
1944	Ingrid Bergman	29	Gaslight
1945	Joan Crawford	41	Mildred Pierce
1946	Olivia de Havilland	30	To Each His Own
1947	Loretta Young	34	The Farmer's Daughter
1948	Jane Wyman	34	Johnny Belinda
1949	Olivia de Havilland	33	The Heiress
1950	Judy Holliday	28	Born Yesterday
1951	Vivien Leigh	38	Streetcar named Desire
1952	Shirley Booth	45	Come Back Little Sheba
1953	Audrey Hepburn	24	Roman Holiday
1954	Grace Kelly	26	The Country Girl
1955	Anna Magnani	48	The Rose Tattoo
1956	Ingrid Bergman	41	Anastasia
1957	Joanne Woodward	27	The Three Faces of Eve
1958	Susan Hayward	40	I Want to Live
1959	Simone Signoret	38	Room at the Top
1960	Elizabeth Taylor	28	Butterfield 8
1961	Sophia Loren	27	Two Women
1962	Anne Bancroft	31	The Miracle Worker
1963	Patricia Neal	37	Hud
1964	Julie Andrews	30	Mary Poppins
1965	Julie Christie	24	Darling
1966	Elizabeth Taylor	34	Who's Afraid of Virginia Woolf?

Year	Actress	Age	Movie
1967	Katharine Hepburn	60	Guess Who's Coming to Dinner?
1968	Katharine Hepburn	61	The Lion in Winter
double award	Barbara Streisand	26	Funny Girl
1969	Maggie Smith	35	The Prime of Miss Jean Brodie
1970	Glenda Jackson	34	Women in Love
1971	Jane Fonda	34	Klute
1972	Liza Minelli	26	Cabaret
1973	Glenda Jackson	37	A Touch of Class
1974	Ellen Burstyn	42	Alice Doesn't Live Here Anymore
1975	Louise Fletcher	41	One Flew Over the Cuckoo's Nest
1976	Faye Dunaway	35	Network
1977	Diane Keaton	31	Annie Hall
1978	Jane Fonda	41	Coming Home
1979	Sally Field	33	Norma Rae
1980	Sissy Spacek	30	Coal Miner's Daughter
1981	Katharine Hepburn	74	On Golden Pond
1982	Meryl Streep	33	Sophie's Choice
1983	Shirley MacLaine	49	Terms of Endearment
1984	Sally Field	38	Places in the Heart
1985	Geraldine Page	61	Trip to Bountiful
1986	Marlee Matlin	21	Children of a Lesser God
1987	Cher	41	Moonstruck
1988	Jodie Foster	26	The Accused
1989	Jessica Tandy	80	Driving Miss Daisy
1990	Kathy Bates	42	Misery
1991	Jodie Foster	29	The Silence of the Lambs

[1]This exploration is based on an article by Richard Brown and Gretchen Davis, published in *Mathematics Teacher*, February, 1990 issue.

Copyright © 1998 by Brendan Kelly Publishing Inc.

"BEST ACTORS" DURING A HALF CENTURY— 1942-1991

Year	Actor	Age	Movie
1942	James Cagney	43	Yankee Doodle Dandy
1943	Paul Lukas	48	On the Rhine
1944	Bing Crosby	43	Going My Way
1945	Ray Milland	40	The Lost Weekend
1946	Fredric March	49	Best Years of our Lives
1947	Ronald Colman	56	A Double Life
1948	Laurence Olivier	41	Hamlet
1949	Broderick Crawford	38	All the King's Men
1950	Jose Ferrer	38	Cyrano de Bergerac
1951	Humphrey Bogart	52	The African Queen
1952	Gary Cooper	51	High Noon
1953	William Holden	35	Stalag 17
1954	Marlon Brando	30	On the Waterfront
1955	Ernest Borgnine	38	Marty
1956	Yul Brynner	41	The King and I
1957	Alex Guinness	43	Bridge on the River Kwai
1958	David Niven	49	Separate Tables
1959	Charleton Heston	35	Ben Hur
1960	Burt Lancaster	47	Elmer Gantry
1961	Maximillian Schell	31	Judgment at Nuremburg
1962	Gregory Peck	46	To Kill a Mockingbird
1963	Sidney Poitier	39	Lillies of the Field
1964	Rex Harrison	56	My Fair Lady
1965	Lee Marvin	41	Cat Ballou
1966	Paul Scofield	44	A Man for All Seasons

Year	Actor	Age	Movie
1967	Rod Steiger	42	In the Heat of the Night
1968	Cliff Robertson	43	Charly
1969	John Wayne	62	True Grit
1970	George C. Scott	43	Patton
1971	Gene Hackman	40	The French Connection
1972	Marlon Brando	48	The Godfather
1973	Jack Lemmon	48	Save the Tiger
1974	Art Carney	56	Harry and Tonto
1975	Jack Nicholson	38	One Flew Over the Cuckoo's Nest
1976	Peter Finch	60	Network
1977	Richard Dreyfuss	32	The Goodbye Girl
1978	Jon Voight	40	Coming Home
1979	Dustin Hoffman	42	Kramer vs. Kramer
1980	Robert de Niro	37	Raging Bull
1981	Henry Fonda	76	On Golden Pond
1982	Ben Kingsley	39	Gandhi
1983	Robert Duvall	55	Tender Mercies
1984	F. Murray Abraham	45	Amadeus
1985	William Hurt	35	Kiss of the Spider Woman
1986	Paul Newman	61	Color of Money
1987	Michael Douglas	33	Wall Street
1988	Dustin Hoffman	51	Rainman
1989	Daniel Day Lewis	32	My Left Foot
1990	Jeremy Irons	42	Reversal of Fortune
1991	Anthony Hopkins	54	The Silence of the Lambs

WORKED EXAMPLE

Display the ages of the "best actress" and the "best actor" winners in a back-to-back stem-and-leaf plot. Create and trace the box plots for these ages to find the age quartiles for "best" actors and actresses.

SOLUTION

To create a stem-and-leaf plot, for the ages of the actresses and actors, enter the ages of "best actresses" and "best actors" into lists, L_1 and L_2 respectively. To sort L_1 in ascending order, press these keys: 2nd [LIST] ▶ ENTER 2nd [L_1] ENTER

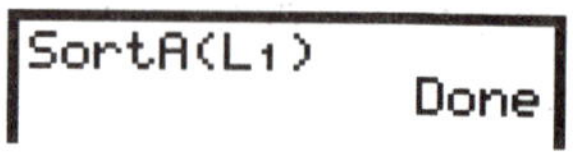

Similarly, we sort L_2. The lists L_1 and L_2 on the STAT Edit menu yield the following.

Back-to-Back Stem-and-Leaf Plot

Ages of "Best Actresses"	Stem	Ages of "Best Actors"
9 9 8 8 7 7 6 6 6 6 4 4 4 1	2	
8 8 8 7 7 5 5 4 4 4 4 4 4 3 3 3 1 1 0 0 0	3	0 1 2 2 3 5 5 7 8 8 8 8 9 9
9 8 5 2 2 1 1 1 1 1 0	4	0 0 0 1 1 1 2 2 2 3 3 3 3 3 4 5 5 6 7 8 8 8 9 9
	5	1 1 2 4 5 6 6 6
1 1 0	6	0 1 2
4	7	6
0	8	

This represents 35. We see it occurs twice.

This represents 49

To create a box plot, for the ages of the actresses and actors, follow these steps:

1. Press: 2nd [STAT PLOT]. Then define Plot 1 and Plot 2 as shown below.

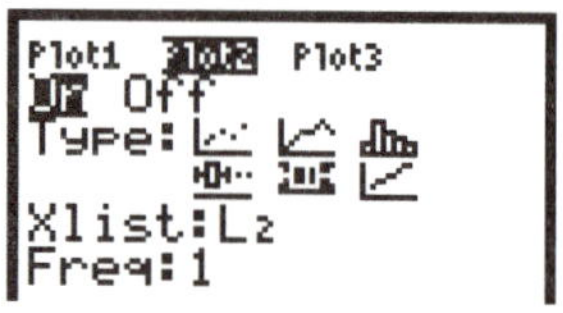

2. Press: ZOOM 9 to obtain these box plots.

3. Trace along the box plots to find the quartiles in the display.

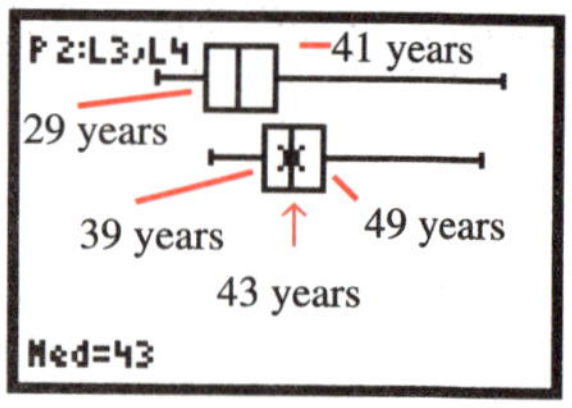

Copyright © 1998 by Brendan Kelly Publishing Inc.

Exercises

1. Follow the procedures in the *Worked Example* to create the box plots showing the age distributions of "best actress" and "best actor" award winners.

a) Compare the median age of the "best actor" award winners with the median age of "best actress" award winners.

b) Why might you expect the "best actress" award winners to have a younger median age than the "best actor" award winners?

c) Trace along your box plots to find the lower and upper extreme (minimum and maximum) ages of the award winners.

(Note: To move from one box plot to the other, press: ▼)

d) Compare the lower extreme age of the "best actresses" with the lower extreme age of the "best actors". Why might you expect the lower extreme age of the "best actress" winners to be less than the lower extreme age of the "best actor" winners?

e) Compare the upper extreme age of the "best actresses" with the upper extreme age of the "best actors" . Why might you expect the upper extreme age of the "best actress" winners to be close to the upper extreme age of the "best actor" winners?

f) The difference between the upper and lower extremes is called the *range*. Calculate and compare the ranges for the "best actors" and for the "best actresses".

g) The difference between the first quartile Q_1 and the third quartile Q_3 is called the *interquartile range*. Calculate and compare the interquartile ranges for the "best actors" and for the "best actresses".

h) Which difference do you think will be greater: the differences in the median ages of the "best actor" and "best actresses", or the difference in their means? Explain.
Access the **1-Var Stats** using the keying sequence described in the *Worked Example* on page 13 to find the mean ages of the "best actress" and "best actor" award winners.

2. Box plots like the one in the *Worked Example* are sometimes called *box-and-whisker* plots. The box is the rectangular region bounded by Q_1 and Q_3 and the whiskers are the horizontal lines extending from the box to the extremes.

a) What percent of the data are represented by values inside the box?

b) What percent of the data are represented by values between and including the ends of the whiskers?

c) When does the median occur at the center of the box?

d) Describe a data set which has a narrow box and long whiskers?

Investigations

3. Is there a trend toward selecting older women for the "best actress" award?

a) Construct a box plot for the ages of the "best actresses" spanning the quarter century from 1942 to 1966.

b) On the same display, construct a box plot for the ages of the "best actresses" for the quarter century from 1967 to 1991.

c) Compare the box plots in parts a) and b). Does there appear to be a trend toward selecting older women? Explain.

d) Construct box plots for the ages of the "best actors"over the two quarter centuries 1942-66 and 1967-91. Is there a trend toward selecting older men for the "best actor" award?

4. a) The *mode* in a set of data is the value that occurs most often. What is the modal age for the "best actress" award winners over the period 1942-91?

b) What is the modal age for the "best actor" award winners over the period 1942-91?

5. We noted in exercise 1g) that the length of the box in the box plot of any set of data is called the *interquartile range*. Any value in a set of data whose distance from the box exceeds 1.5 times the interquartile range is called an *outlier*.

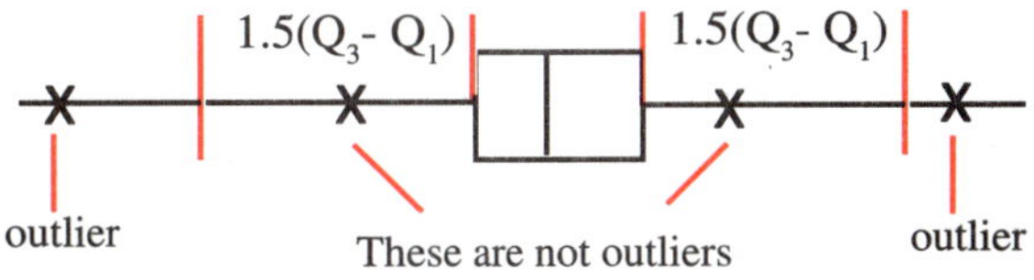

a) List any outliers in the set of ages of "best actress" winners.
b) List any outliers in the set of ages of "best actor" winners.
c) Construct ModBoxPlots to verify your answers in a) & b).
d) Explain the significance of an outlier in the set of ages of award winners.

CHALLENGE

In a set of data, $\{x_1, x_2, x_3, \ldots x_n\}$, the deviation of x_i from the average is given by $|x_i - x_{average}|$. Which value for $x_{average}$, the mean or the median, minimizes the sum of the deviations for all x_i ?

To calculate the sum of the deviations for the ages of the actresses using $x_{average} = \bar{x}$, define L_5 by abs($L_1 - \bar{x}$)->L_5 and then L_5 X L_2 -> L_6. Then evaluate **sum** L_6. Repeat the calculation replacing $\bar{x}$ with Med to evaluate the sum of the deviations $|x_i$ - Med$|$.

Copyright © 1998 by Brendan Kelly Publishing Inc.

EXPLORATION 4 Measures of Dispersion

Addison-Wesley Canada © 1991
Reprinted with Permission

In Explorations 2 and 3, we studied two different measures of central tendency, the mean and the median. If we imagine a set of numbers $x_1 < x_2 < x_3 < \ldots < x_6$ distributed along a number line, the *mean* would correspond to the center of gravity or balance point of the set.

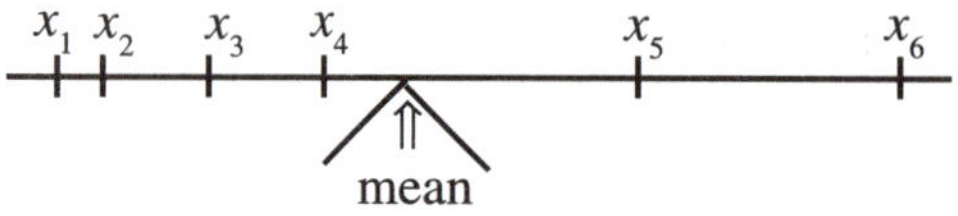

That is, the sum of the differences $x_i - \bar{x}$ is 0. Adding any two data points to our set does not change the mean if the two new values are equidistant from the mean and on opposite sides.

Similarly, the *median* in this set corresponds to the middle value (or the arithmetic mean of the two middle values). Adding two new data points to the set does not change the median provided the two new points are on opposite sides of the median. In short, measures of central tendency tell us the "midpoint" of a set of data but they do not tell us how closely the values cluster around this midpoint. An old adage says:

If your head is in the furnace and you're standing on
a block of ice, your average body temperature is normal.

In the ***CHALLENGE*** investigation of Exploration 3, you may have touched upon the following interesting properties of the mean and the median.

> Consider all the distances $|x_i - x_0|$ between the points in the data set and a particular point, x_0. The sum of these distances is a minimum when x_0 is the median and the sum of the squares of these distances is a minimum when x_0 is the mean

The quartiles and the corresponding box plots studied in the previous explorations provide some measure of the distribution of the data around the median. To provide another measure of the extent to which the data "cluster" around the mean, statisticians have developed measures of *dispersion.* The most widely used measure of dispersion is the *standard deviation* (denoted by the Greek letter σ (sigma) for populations and the Roman letter s for samples). The (sample) standard deviation of the set of data $\{x_1, x_2, x_3, \ldots x_n\}$ is defined by

$$s = \sqrt{\frac{(x_1-\bar{x})^2+(x_2-\bar{x})^2+\ldots\ldots\ldots+(x_n-\bar{x})^2}{n-1}}$$

Although this formula appears complicated, it is composed as follows: We square all the *deviations from the mean* $|x_i - \bar{x}|$ (i. e. we square the "distance" of each x_i from the mean) and divide by one less than the number of data. Then take the square root to get s. The squaring of the deviations from the mean makes s more sensitive to extreme values in the data. If the gentleman in the cartoon had insisted on batteries whose lifetimes have a small standard deviation, his watch would probably have run for close to 5.7 years.

Copyright © 1998 by Brendan Kelly Publishing Inc.

Worked Example

The lifetimes (in years) of 30 Brand A and 30 Brand B batteries are given in the tables. Which of these brands of battery is more reliable?

Measured Lifetimes of 30 Brand A Batteries				
5.1	7.3	6.9	4.7	4.6
6.2	6.4	5.5	4.9	6.9
6.0	4.8	4.1	5.3	8.1
6.3	7.5	5.0	5.7	9.3
3.3	3.1	4.3	5.9	6.6
5.8	5.0	6.1	4.6	5.7

Measured Lifetimes of 30 Brand B Batteries				
5.4	6.3	5.0	5.9	5.6
4.7	6.0	3.3	6.6	6.0
5.0	6.5	5.8	5.4	4.9
5.7	6.8	5.6	4.9	6.0
4.9	5.7	6.2	7.5	5.8
6.8	5.9	5.3	5.6	5.9

Solution

To compare the distributions of the lifetimes of the two brands we may proceed as in *Exploration* 3 p. 16 and create a back-to-back stem-and-leaf plot for the lifetimes of each brand.

Back-to-Back Stem-and-Leaf Plot

Brand A	Stem	Brand B
3 1	3	3
9 8 7 6 6 3 1	4	7 9 9 9
9 8 7 7 5 3 1 0 0	5	0 0 3 4 4 6 6 6 7 7 8 8 9 9 9
9 9 6 4 3 2 1 0	6	0 0 0 2 3 5 6 8 8
5 3	7	5
1	8	
3	9	

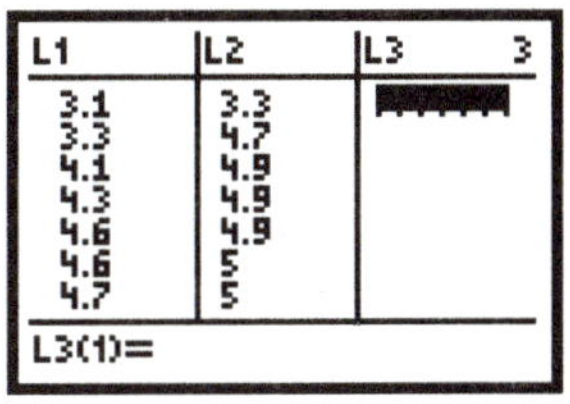

To create box plots, for the lifetimes of brands A and B, we follow these steps:

❶ Press: 2nd [STAT PLOT]. Then define Plot 1 and Plot 2 as shown below.

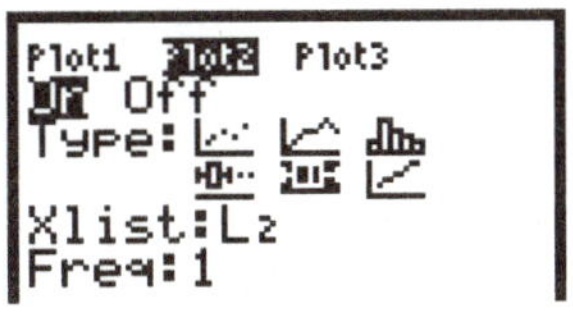

❷ Press: ZOOM 9 to obtain these box plots.

❸ Trace along the box plots to find the quartiles in the display.

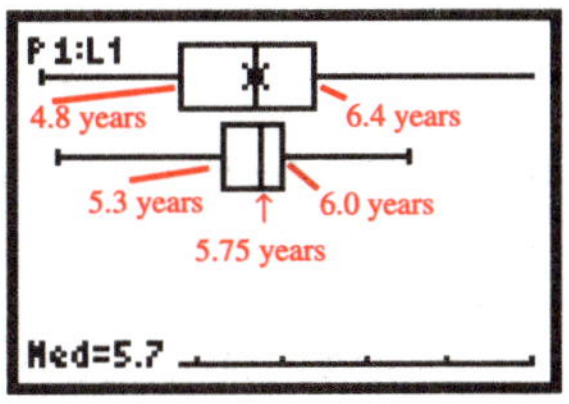

We observe from the box plots that both batteries have about the same median. However, the interquartile range (length of the box) for Brand A batteries (upper box plot in the display) is much greater than the interquartile range for the Brand B batteries. Therefore, the Brand B batteries are more reliable. We can also compare reliabilies by computing $\bar{x}$ and s. We use s (shown on your TI-83 display as Sx) as an estimator of the population standard deviation σ_x. The definition of σ_x the standard deviation for populations is the same as for samples except we replace $\bar{x}$ with the population mean, μ and replace $n-1$ with n. (We deal with this in more detail later– see *Investigation* 8 p. 71.)

To obtain $\bar{x}$ and s, we press: STAT ▶ ENTER to access **1-Var Stats**.

We then enter L_1 and press ENTER.

Brand A

Brand B

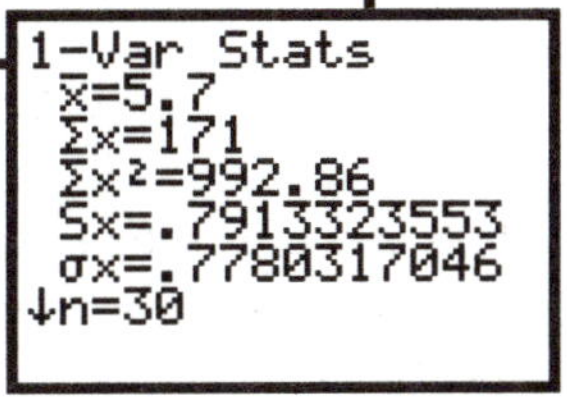

We obtain the 1-variable statistics display for Brand A. Similarly, we obtain the display for Brand B. Both brands of battery have a mean life of 5.7 years; however, the estimate Sx of the standard deviation, σ_x is about 1.35 years for Brand A, and only about 0.79 for Brand B, so Brand B is more reliable.

Copyright © 1998 by Brendan Kelly Publishing Inc.

Exercises

1. Why did statisticians develop measures of dispersion?

2. The simplest measure of dispersion is called the *range*. The *range* of a set of data is defined to be the difference between the largest and smallest values. In the worked example:

a) What is the range of the lifetimes of the Brand A batteries?
b) What is the range of the lifetimes of the Brand B batteries?
c) Why is the standard deviation used more often than the range as a measure of dispersion?

3. Another measure of dispersion, called the *mean deviation* is designed to take into account the deviation of each number in a set of data from the mean. The *mean deviation* is defined to be the mean of all the "distances" between the x_i and $\bar{x}$. That is, the mean deviation of a set of data, $\{x_1, x_2, x_3, \ldots x_n\}$ is:

$$\frac{|x_1 - \bar{x}| + |x_2 - \bar{x}| + |x_3 - \bar{x}| + \ldots + |x_n - \bar{x}|}{n}$$

(Vertical bars denote absolute value.)

a) Determine the mean deviation in the lifetimes of Brand A batteries. (Do the subtractions in your head and enter the values $|x_i - \bar{x}|$ into the calculator to form a running total.)

b) Compute the mean deviation for the Brand B batteries.

c) Compare your answers in parts (a) and (b) and determine which brand of battery is more reliable according to this measure of dispersion.

4. Calculate the mean deviation and the standard deviation for
a) S = {1, 2, 3, 4, 5, 6, 7, 8, 9, 10, 11}
b) T = {-5, -4, -3, -2, -1, 0, 1, 2, 3, 4, 5}
What is the effect on the standard deviation of a set of data when a fixed constant is added to all the values?

5. Compare the mean deviation (see Exercise 3) and the standard deviation as measures of dispersion. Which measure is more sensitive to data which has some large deviations from the mean?

6. The reputation of a steel processing plant rests on its ability to manufacture products within fine tolerance limits. Samples of 30 ball bearings which are advertised as 8 mm in diameter by Acme Bearings were measured and compared with a sample of 30 produced by Delta Bearings. Calculate $\bar{x}$ and s for the samples below, to determine which company produces bearings with the smaller deviations from 8 mm.

Diameters of 30 Ball Bearings from Acme				
7.93	7.87	7.84	8.09	8.14
8.28	7.88	8.01	7.92	7.98
8.01	8.05	8.01	8.07	7.94
8.03	8.14	8.09	7.98	8.01
7.93	7.98	8.09	7.95	8.04
8.13	7.90	7.89	8.18	8.13

Diameters of 30 Ball Bearings from Delta				
8.01	8.04	8.09	7.98	8.05
8.10	8.05	8.09	7.99	7.99
7.97	7.99	8.00	8.03	8.03
7.94	7.97	8.03	7.93	8.10
8.00	8.03	8.06	8.08	8.11
8.14	7.96	8.05	8.08	8.00

Investigations

7. One of the histograms below displays the lifetimes of the 30 Brand A batteries and the other displays the lifetimes of the 30 Brand B batteries presented in the *Worked Example*. (With window settings: $0 \le x \le 10$, $0 \le y \le 16$). Plot the histogram of the lifetimes of either brand using these settings and identify the matching histogram. Use class intervals of 1.

a) 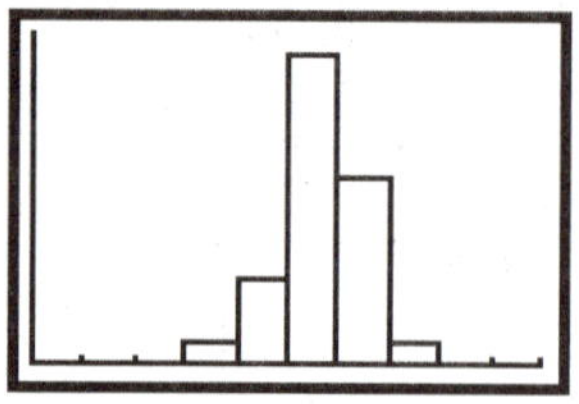

b) 

Double the class intervals in your histogram. What happens to the shape of the histogram when the class interval doubles? Why?

8. Describe how the shape of a histogram is related to the dispersion of a set of data.

9. Construct box plots for the diameters of the ball bearings manufactured by each of the two companies in Exercise 6. Use the shapes of the box plots to determine which set of ball bearings has the smaller dispersion. Compare this conclusion with your answer to Exercise 6.

10. Create (if possible) two sets of data which have the same mean, so that one of the sets has a greater standard deviation but a smaller interquartile range than the other set.

11. The vertical line with equation $x = a$ divides the area of a histogram in half. Is a the value of the mean? the median?

12. The variables z_i are related to variables x_i by the equation $z_i = \frac{x_i - \bar{x}}{\sigma}$ where σ is the population standard deviation of the x_i. Calculate the mean and standard deviation for the z_i.

CHALLENGE

The symbol Σ (upper case sigma) is used to denote a sum.

To write $x_1 + x_2 + x_3 + \ldots + x_n$ we write: $\sum_1^n x_i$

Write the definition of $\bar{x}$ and s using the sigma notation.

Enter the natural numbers from 1 to 10 into L_1 of the STAT edit table.

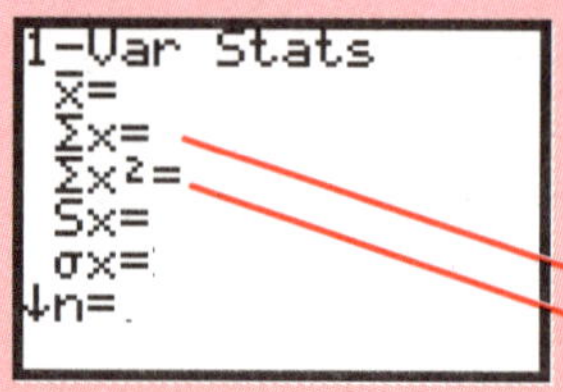

Press STAT ▶ ENTER to obtain **1-Var Stats** and use this command to find the sum of these numbers and the sum of their squares.

Copyright © 1998 by Brendan Kelly Publishing Inc.

UNIT 2 TWO VARIABLE STATISTICS

THE STEPS IN ANALYSING & PLOTTING TWO-VARIABLE DATA

STEP 1 ENTER YOUR DATA INTO A LIST

To enter your data into a list, press: **STAT**

We get this display… then press **ENTER** to obtain

```
EDIT CALC TESTS
1:Edit…
2:SortA(
3:SortD(
4:ClrList
5:SetUpEditor
```

```
L1      L2      L3     1
------  ------  ------



L1(1) =
```

Then enter your data into lists in the STAT matrix.

IF YOU WANT ONLY REGRESSION STATISTICS, SKIP TO STEP 3.

STEP 2 OBTAIN TWO VARIABLE STATISTICS

To obtain statistics involving two data variables, press: **STAT** **▶**

We obtain this display… and then press **2** to obtain:

```
EDIT CALC TESTS
1:1-Var Stats
2:2-Var Stats
3:Med-Med
4:LinReg(ax+b)
5:QuadReg
6:CubicReg
7↓QuartReg
```

```
2-Var Stats
```

Enter the appropriate listnames (e.g. L_1, L_2) .

```
2-Var Stats L1,L
2
```

You get a display like this.

```
2-Var Stats
 x̄=4.11
 Σx=123.3
 Σx²=685.97
 Sx=2.485870416
 σx=2.444088105
 n=30
```

```
2-Var Stats
↑ȳ=77.16666667
 Σy=2315
 Σy²=186161
 Sy=16.10329444
 σy=15.83263156
 Σxy=10402.8
 minX=.5
 maxX=9.3
 minY=40
 maxY=100
```

STEP 3 TO GET REGRESSION STATS

To obtain regression statistics, press: **STAT** **▶**

and enter the number of the regression curve of interest .

```
EDIT CALC TESTS
1:1-Var Stats
2:2-Var Stats
3:Med-Med
4:LinReg(ax+b)
5:QuadReg
6:CubicReg
7:QuartReg
8:LinReg(a+bx)
9:LnReg
0:ExpReg
A:PwrReg
B:Logistic
C:SinReg
```

You display like this.

```
LinReg(ax+b) █
```

Enter the data lists you wish to compare and the Y variable to which you wish to store the regression equation.

```
LinReg(ax+b) L1,
L2,Y1█
```

Press **ENTER** to obtain a display showing the regression data. You will obtain a display somewhat like this.

```
LinReg
 y=ax+b
 a=4.956000603
 b=56.79750419
 r²=.5853157424
 r=.7650593064
█
```

If neither r^2 nor **r** appears, select the command **DiagnosticOn** from the catalogue menu.

STEP 4 DISPLAY THE DATA PLOT.

Define a data plot by selecting the appropriate type of graph and defining the **Xlist** and **Ylist**.

```
Plot1 Plot2 Plot3
On Off
Type: [scatter] [xyLine] [histogram]
      [modified box] [box] [normal prob]
Xlist:L1
Ylist:L2
Mark: □ + ·
```

To display the data plot, press **ZOOM** **9**

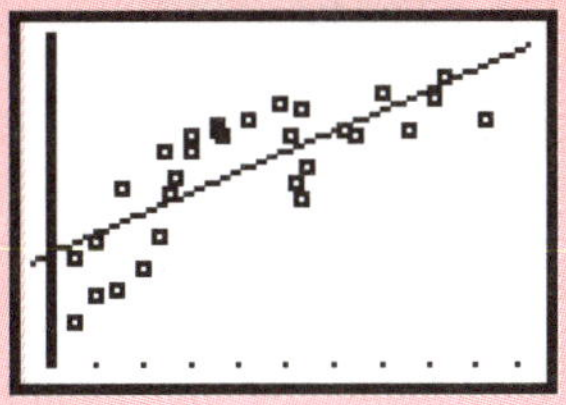

Copyright © 1998 by Brendan Kelly Publishing Inc.

EXPLORATION 5 Scatterplots & Correlation

Addison-Wesley Canada © 1991
Reprinted with Permission

The students in Mr. Santos's class were asked to record the number of hours spent studying for their mathematics test. For each student, Mr. Santos wrote an ordered pair (x, y). The value of x represented the number of hours of study and the value of y was the student's mark on the test. Marsha's ordered pair was (3, 82) because she spent 3 hours studying and received a final mark of 82 out of 100. The set of ordered pairs that Mr. Santos recorded are shown on the right.

Ordered Pairs

(3.0, 82)	(5.5, 78)	(1.0, 60)	(4.9, 93)	(5.1, 86)
(2.5, 71)	(4.2, 90)	(0.5, 40)	(3.5, 88)	(7.0, 96)
(1.5, 73)	(2.4, 82)	(2.0, 53)	(6.2, 87)	(8.4, 100)
(2.6, 75)	(3.7, 85)	(5.4, 70)	(9.3, 89)	(7.6, 87)
(1.4, 48)	(0.5, 56)	(6.5, 85)	(2.3, 61)	(5.2, 74)
(1.0, 47)	(3.5, 87)	(8.2, 94)	(3.0, 86)	(5.4, 92)

WORKED EXAMPLE 1

Use the data given above to determine whether there is a relationship between the number of hours of study and the test mark.

SOLUTION

To enter these data, we access the STAT EDIT table by pressing: STAT ENTER

We clear the table and enter in L_1 the first coordinates of all the ordered pairs and in L_2, all the second coordinates, taking care to check that opposite each value in L_1 is the corresponding second coordinate in L_2.

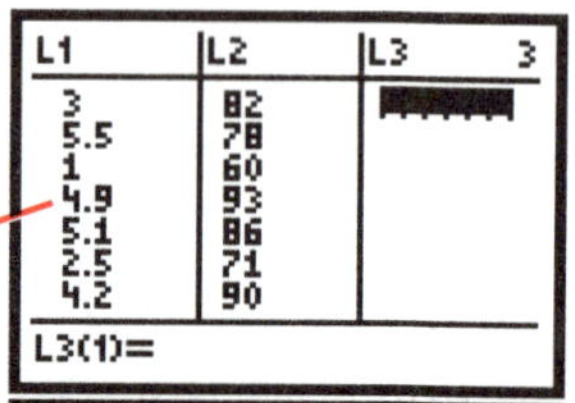

Row 4 corresponds to the ordered pair (4.9, 93)

The display shows the first 7 rows of the table.

To determine whether there is a relationship between the number of hours of study and the test mark we plot the 30 data points corresponding to the 30 ordered pairs. To do this, we first turn off all statistical plots by entering:

2nd [STAT PLOT] 4 ENTER followed by 2nd [STAT PLOT] ENTER

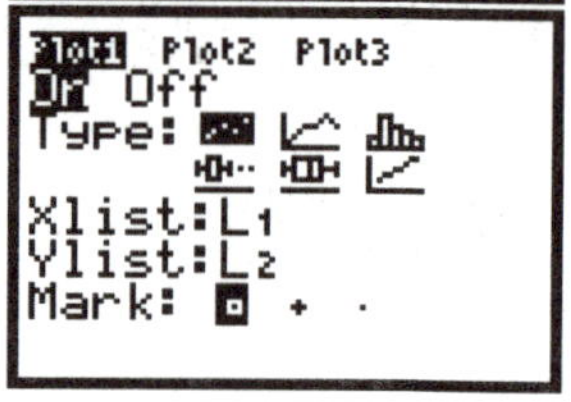

We define Plot 1 by selecting the options as shown in the display. That is, we choose the first icon in row "Type", which represents a type of plot called a *scatterplot.* We enter L_1 as the Xlist and L_2 as the Ylist.

To graph this plot, we press: ZOOM 9 to get the display shown here.

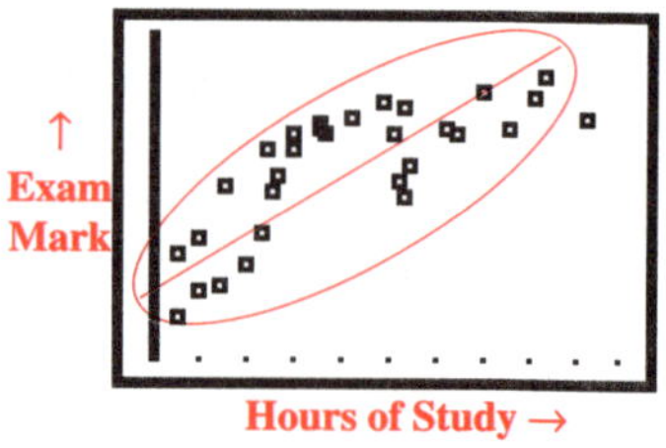

The display shows that almost all the data points in the scatterplot can be contained in an oval which is inclined along an axis with positive slope. In such a case we say that the variables plotted on both axes are *positively correlated.* The scatterplot shows that there is a positive correlation between preparation time for a test and the test mark. When the data points cluster close to the axis of the oval, we say the variables have a *strong positive correlation.* This scatterplot shows a fairly strong positive correlation.

Copyright © 1998 by Brendan Kelly Publishing Inc.

IS A CAR AN INVESTMENT OR AN EXPENDITURE?

When an item that you purchase tends to increase in value, we usually refer to it as an *investment*. Conversely, an item that decreases in value, is called an *expenditure*. Understanding the difference between an investment and an expenditure is an important key to managing your personal finances. To determine whether a car is an expenditure or an investment, we must find out whether its value increases or decreases with its age.

The table displayed here shows for a popular model car, the resale value in dollars for each age between 1 and 10 years.

Age	Value	Age	Value
1	$15,975	6	$3,448
2	$ 9,285	7	$2,755
3	$ 8,000	8	$1,995
4	$ 6,790	9	$1,400
5	$ 3,150	10	$1,268

WORKED EXAMPLE 2

Use the data given in the table to determine whether there is a correlation between the age of a car and its resale value.

SOLUTION

By scanning the table, we see that the resale value of the car tends to decrease with increasing age, so there seems to be a relationship, i.e. a *correlation*, between these two variables. To determine whether it is a strong or a weak correlation, we proceed as in *Worked Example* 1.

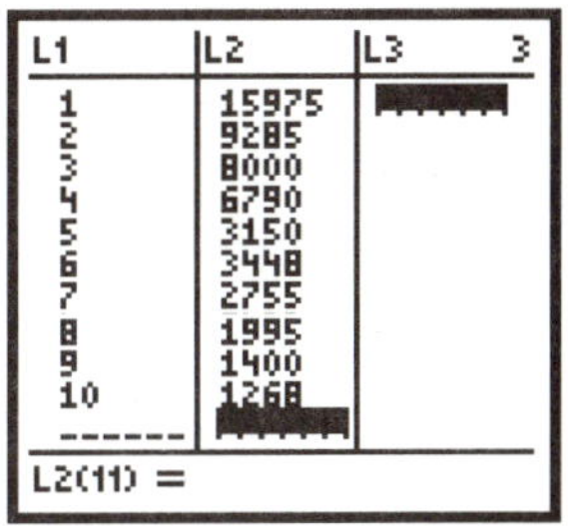

To enter these data, we access the STAT EDIT table by pressing: STAT ENTER
We clear the table and enter in L_1 the ages from 1 to 10 years and in L_2 all the corresponding resale values.

To determine whether there is a relationship between the age of that particular model car and the resale value, we plot the 10 data points as 10 ordered pairs. To do this, we first turn off all statistical plots by entering:

2nd [STAT PLOT] 4 ENTER followed by 2nd [STAT PLOT] ENTER

We define Plot 1 choosing the same options as in *Worked Example* 1, and press:
ZOOM 9 to get the display shown here.

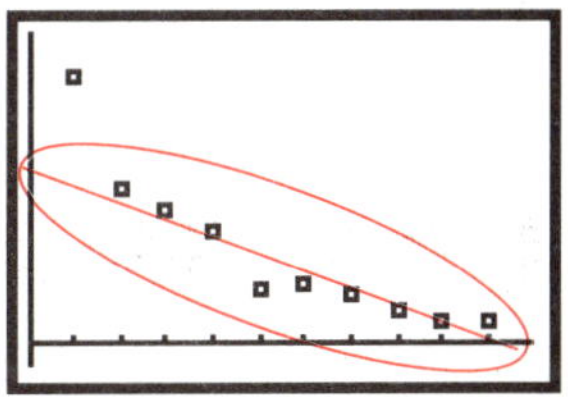

The display shows that almost all the data points in the scatter plot can be contained in an oval that is inclined along an axis with negative slope. In such a case we say that the variables plotted on both axes are *negatively correlated.* The scatterplot shows that there is a negative correlation between the age of a car and its resale value. This scatterplot shows a very strong negative correlation.

Note: It is important to understand that if two variables are strongly correlated, it is not necessarily true that a change in one variable "causes" a change in the other. Often changes in both variables result from a common "cause". For example, shoe size and reading ability of school children are positively correlated because increases in both these variables come with increased age. However, having large feet does not "cause" increased reading ability.

Copyright © 1998 by Brendan Kelly Publishing Inc.

Exercises

1. Write in your own words the meanings of the following phrases:

- •positive correlation
- •strong positive correlation
- •negative correlation
- •strong negative correlation
- •no correlation

2. Use the phrases above to identify the kind of correlation which is characterized by each of the following scatterplots.

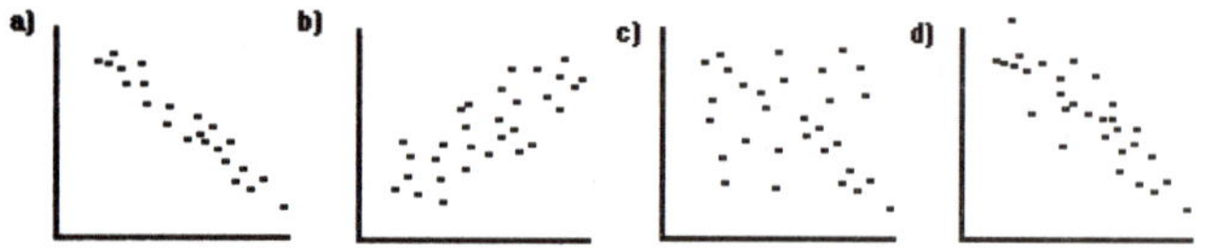

3. The table below shows 30 ordered pairs. The first coordinate in each ordered pair is the age of a randomly selected married woman. The second component is the age of her spouse. Use the procedure in the worked examples to create a scatterplot that will reveal whether there is a correlation between the age of a woman and her spouse.

Ages of Women and their Husbands

(29, 34)	(37, 38)	(19, 20)	(57, 57)	(34, 32)
(23, 25)	(51, 54)	(72, 81)	(29, 23)	(70, 70)
(45, 54)	(39, 37)	(58, 56)	(64, 71)	(35, 35)
(42, 50)	(25, 24)	(37, 48)	(36, 36)	(27, 32)
(56, 42)	(17, 24)	(28, 28)	(57, 26)	(41, 39)
(47, 47)	(16, 18)	(24, 27)	(84, 87)	(55, 59)

If you discover a correlation, indicate whether it is positive or negative and describe it as strong or weak.

4. A biologist who was studying crickets hypothesized that the number of chirps per minute made by a cricket is strongly correlated to the outside temperature increases. He recorded the data shown in the table. Create a scatterplot to test his hypothesis.

Temperature in °C	Chirps per min
17	105
18	110
19	110
20	126
21	126
22	130
23	130
24	152
24	156
25	160
26	170
27	171
28	175
29	196
30	212

5. For each of the following pairs of variables, indicate what kind of correlation (if any) you would expect to exist. Explain why.

a) height and weight

b) age and personal savings

c) level of education and income

d) latitude of a U. S. city and its mean temperature in winter

e) A golfer's age and golf score

f) travel time and speed of travel for a particular distance

g) I. Q. and number of olives eaten per month

Investigations

6. In order to quantify the "degree" of correlation between two variables, Karl Pearson, one of the pioneers of statistics, defined the *correlation coefficient*, *r*, between two variables *x* and *y* by the equation:

$$r = \frac{\sum_{i=1}^{n}(x_i - \bar{x})(y_i - \bar{y})}{\sqrt{\sum_{i=1}^{n}(x_i - \bar{x})^2 \sum_{i=1}^{n}(y_i - \bar{y})^2}}$$

where y_i is the value of *y* corresponding to $x = x_i$. $i = 1 \ldots n$

a) Suppose that *x* and *y* are so strongly correlated that *y* is a linear function of x; i.e. $y = ax + b$ for some constants *a, b*. Prove algebraically that $|r| = 1$. How are *x* and *y* related when $r = 1$? How are they related when $r = -1$?

b) What range of values of *r* would indicate:
a strong correlation? a weak correlation? no correlation?

c) Enter into your STAT EDIT table the temperatures and corresponding chirps per minute given in Exercise 4.
To find the correlation coefficient *r* for cricket chirps per minute and temperature, select the command **DiagnosticOn** by pressing **2nd** **CATALOG** and moving the cursor through the catalog menu to this command. Then press **ENTER** twice.
To obtain both *r* and r^2 press:

STAT **▶** **4** **ENTER**

Record the values of *r* and r^2 shown on the display.

7. a) Calculate the correlation coefficient between the ages of a woman and her spouse (See Exercises 3 and 6.)

b) Calculate the correlation coefficient between hours of study and test mark presented in *Worked Example* 1.

DRAWING INFERENCES

A common error is the assumption that if two variables have a high correlation, then one of the variables is the "cause" of the other. This recent newspaper article revealed an interesting trend that resulted from such a false inference. Give a possible reason for the correlation.

Latin's Remarkable Resurrection

(New York)- A real resurrection of Latin is taking place. The upsurge is at the high school level and is very much a result of student demand.... Why are students flocking to Latin, and what do they expect to get out of it?
...There is, allegedly, a clear correlation between taking Latin at school and doing well on scholastic aptitude exams, and the students are well aware, as they always are, of the statistics. They may not see why Latin, in the high-tech age should help them reach their goal, but they are clear on the fact that it does, and plausibly invoking the national adage, "If it works, don't fix it," queue up and sign on.

Copyright © 1998 by Brendan Kelly Publishing Inc.

WILL FEMALES OUTRUN MALES IN THE 21ST CENTURY?

The women's 100-m dash was not an event in the Summer Olympic Games until 1928. In that year, Elizabeth Robinson of the U. S. won the event with a time of 12.2 seconds. Since that time, the women's record for the 100-m dash has decreased substantially.

The table below shows the winning times for the 100-m dash at the Summer Olympic games between 1928 and 1996 for both men and women. Perusal of the table reveals that the winning times seem to be decreasing more quickly for women than for men. Will women eventually outrun the men?

REUTERS/ Tom Szlukovenyi / Archive Photos

Atlanta, 27 July 1996 — Runners Gwen Torrence of the United States (left), the bronze medalist, Marlene Ottey (center) of Jamaica, the silver medalist, and Gail Devers of the United States, the gold medalist, cross the finish line in the women's 100-meter final.

Year	Men's Winning Time (in seconds)		Women's Winning Time (in seconds)	
1928	Percy Williams (Canada)	10.80	Elizabeth Robinson (U.S.)	12.20
1932	Eddie Tolan (U. S.)	10.30	Stanislawa Walasiewicz (Poland)	11.90
1936	Jessie Owens (U. S.)	10.30	Helen Stephens (U. S.)	11.50
1948	Harrison Dillard (U. S.)	10.30	Francina Blanker-Koen (Nethlds)	11.90
1952	Lindy J. Remigino (U. S.)	10.40	Marjorie Jackson (Australia)	11.50
1956	Bobby J. Morrow (U. S.)	10.50	Betty Cuthbert (Australia)	11.50
1960	Armin Hary (Germany)	10.20	Wilma Rudolf (U. S.)	11.00
1964	Robert L. Hayes (U. S.)	10.00	Wyomia Tyus (U. S.)	11.40
1968	James Hines (U. S.)	9.90	Wyomia Tyus (U. S.)	11.00
1972	Valery Borzov (U.S.S.R.)	10.14	Renate Stecher (East Germany)	11.07
1976	Hasely Crawford (Trin. & Tob.)	10.06	Annegret Richter (West Germany)	11.01
1980	Allan Wells (Great Britain)	10.25	Lyudmila Kondratyeva (U.S.S.R.)	11.06
1984	Carl Lewis (U. S.)	9.99	Evelyn Ashford (U. S.)	10.97
1988	Carl Lewis (U. S.)	9.92	Florence Griffith-Joyner (U. S.)	10.54
1992	Linford Christie (Great Britain)	9.96	Gail Devers (U. S.)	10.82
1996	Donovan Bailey (Canada)	9.84	Gail Devers (U. S.)	10.94

A set of measurements, such as winning times in the 100-m dash, taken at regular intervals of time is called a *time series*. By plotting these measurements against time and joining the plotted points with line segments we create an *xyLine*. This *xy*Line enables us to investigate the data for upward or downward trends. In the worked example on page 26 we construct *xy*Lines to investigate the winning times in the 100-m dash for possible trends. We then construct for the women's winning times and the men's winning times a "trend line" called a *median-median line* that smooths out short-term fluctuations.

Copyright © 1998 by Brendan Kelly Publishing Inc.

WORKED EXAMPLE

a) Use the data given in the table on page 25 to construct the *xy*Lines for the Olympic winning times in the 100-m dash for men and for women during the period from 1928 to 1996.

b) Construct the median-median line for the men's and for the women's winning times and trace along these lines to find the point of intersection.

SOLUTION

a) We enter the values from the table into the STAT EDIT table to obtain the table shown in the display, in which L_1 denotes the year, L_2 the men's record time, and L_3 the women's record time.

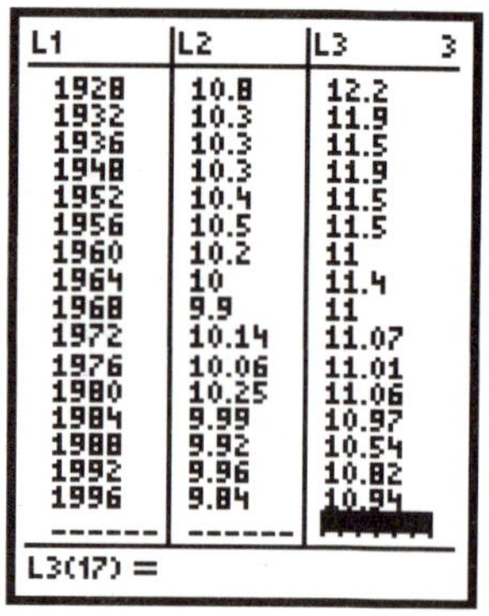

L1	L2	L3 3
1928	10.8	12.2
1932	10.3	11.9
1936	10.3	11.5
1948	10.3	11.9
1952	10.4	11.5
1956	10.5	11.5
1960	10.2	11
1964	10	11.4
1968	9.9	11
1972	10.14	11.07
1976	10.06	11.01
1980	10.25	11.06
1984	9.99	10.97
1988	9.92	10.54
1992	9.96	10.82
1996	9.84	10.94
------	------	

L3(17) =

We define Plots 1 and 2 as shown on the displays. Note that we select the *xy*Line (second icon in "Type") for both plots and different "Marks" for the two plots.

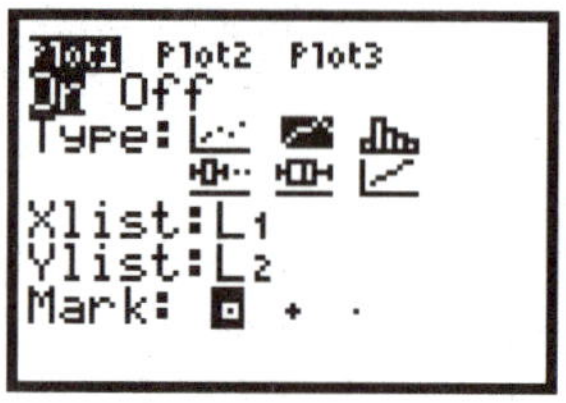

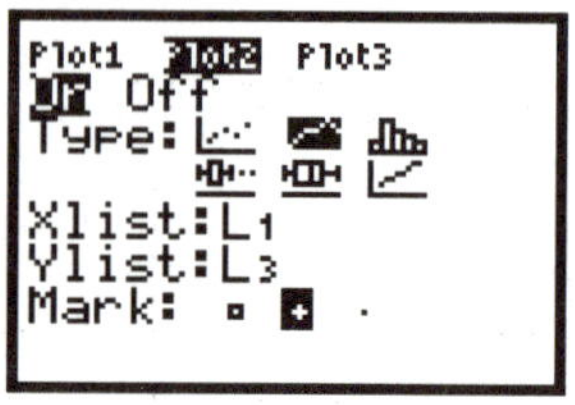

When we activate ZoomStat by pressing: ZOOM 9 , we obtain this display.

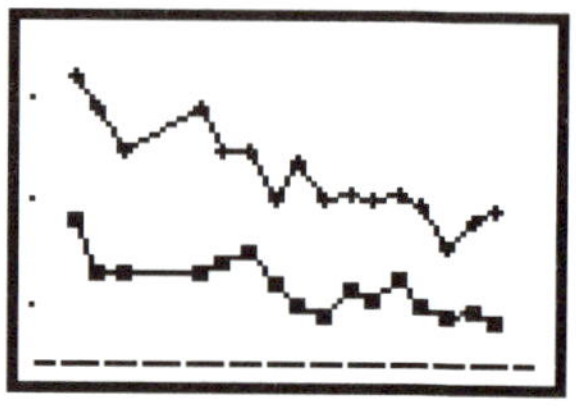

The top *xy*Line in the display is Plot2, representing the women's winning times. This *xy*Line seems to be sloped downward more steeply than the winning times for men.

b) To calculate the median-median line for the men's winning times, we press:

STAT ▶ 3 ENTER and obtain the equation shown in this display.

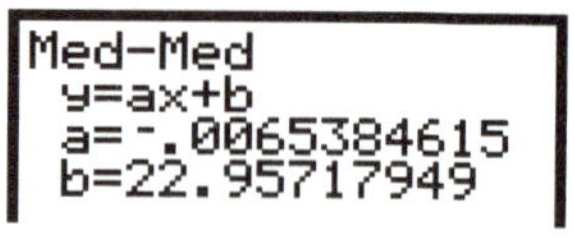

Before plotting the median-median line, we enter it on the "Y = " list by pressing:

Y= VARS 5 ▶ ▶ ENTER

To calculate the median-median line for the women's winning times, we press:

STAT ▶ 3 2nd [L_1] , 2nd [L_3] ENTER

(Note that when no lists are specified, the default Xlist and Ylist are L_1 and L_2.)

This yields the equation shown in the display opposite.
We assign the expression for the median-median line for women's winning times to the variable Y_2 by following the same procedure as for the men's winning times.

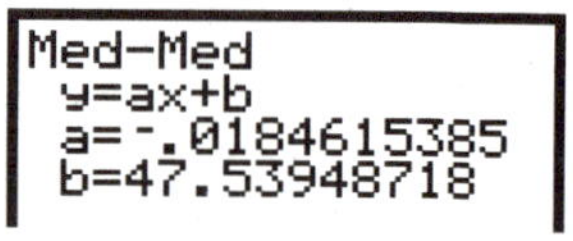

To graph the *xy*Lines together with the median-median lines, we ensure that Plot1 and Plot2 are turned "on" and that the equations of both median lines are selected on the Y= list and then press ZOOM 9 . We get the display shown here.

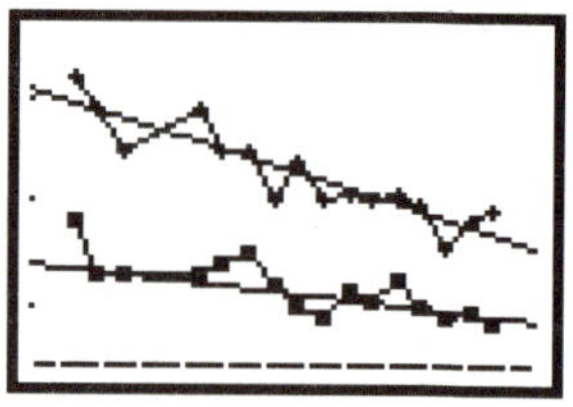

It is clear that the median-median line for the women's winning times has a steeper negative slope than the median-median line for the men's winning times and therefore the two lines will intersect at some point. Since the lines do not intesect in the current window, we redefine the window settings so that $X_{max} = 2100$ and $Y_{min} = 7$.

To find the point of intersection, we press: GRAPH 2nd [CALC] 5

In response to the prompts, **First curve?** and **Second curve?**, we press ENTER

In response to the prompt, **Guess?**, we move the cursor close to the intersection and press ENTER. We obtain this display that suggests that if women and men continue to improve at present rates, around the year 2062 the winning time for both sexes will be about 9.48 seconds. According to this model, what happens after 2062?

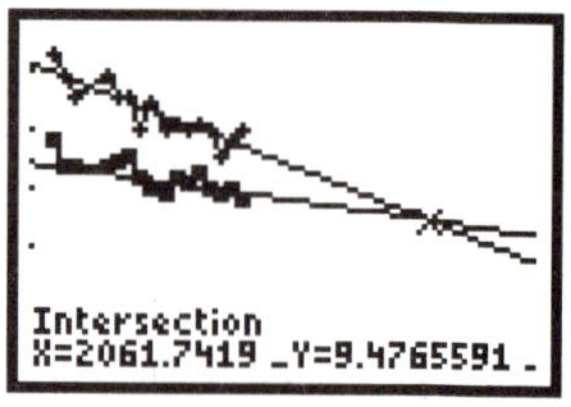

Copyright © 1998 by Brendan Kelly Publishing Inc.

Exercises

1. Write a sentence or two to explain each of the following terms.

- time series
- xyLine
- trend line
- median-median line

2. Do each part of this exercise in sequence to construct a median-median line without using your calculator.

a) Use graph paper to create a grid like the one below.

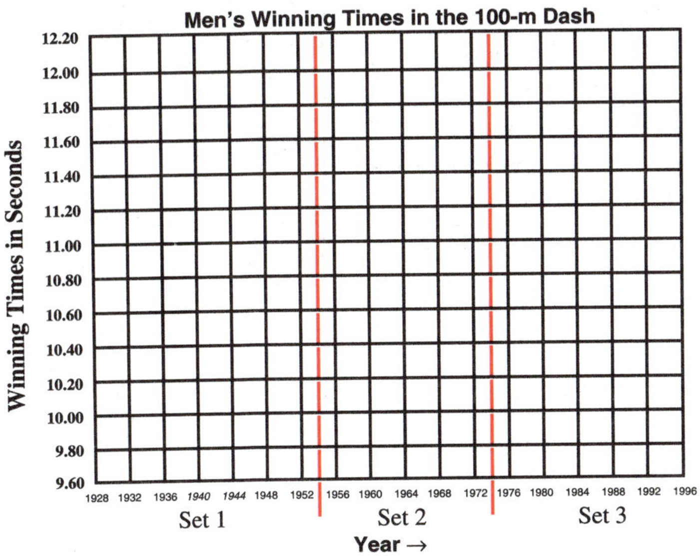

b) Using the table on page 25, plot the points corresponding to the men's winning times.

c) Partition the data into three sets as shown so that there are five consecutive *Olympic years* in each of the first two sets and six consecutive Olympic years in the third set.

d) Identify the median year in set 1 and denote it by x_1, and the median winning time in set 1 and denote it by y_1. Form the ordered pair (x_1, y_1) and call it the *summary point* for set 1.

e) Record the summary points (x_2, y_2) and (x_3, y_3) for set 2 and set 3 respectively.

f) Draw a line through (x_1, y_1) and (x_3, y_3). Draw a second line parallel to the first and through (x_2, y_2). Finally draw a third line between and parallel to the first two so that the distance between the first and third lines is one-third the distance between the first and second lines. This line is the *median-median line.*

g) Measure the slope of the median-median line and its y-intercept and then compare with the equation of the median line for the men's winning times in the *Worked Example*.

Investigations

3. Some scientists believe that increased quantities of carbon dioxide in the earth's atmosphere are causing a gradual global warming that will eventually melt the polar ice caps and cause world-wide flooding. The table below shows the annual number of metric tons of carbon dioxide emissions per capita from fossil fuel consumption and cement manufacture during the period 1950-1989.

Per Capita Carbon Dioxide Emissions in Metric Tons for the period 1950-1989

Year	Emissions	Year	Emissions
1950	2.38	1975	4.14
1955	2.71	1980	4.32
1960	3.15	1985	4.07
1965	3.48	1989	4.21
1970	4.07		

a) Follow the procedures outlined in Exercise 2 to construct the median-median line for the annual number of tons of carbon dioxide emissions per person between 1960 and 1989.

b) Use your TI-83 calculator to calculate the median-median and compare with your answer in part a).

c) If the carbon dioxide emissions continue to increase as in the past, what will be the per capita emissions of carbon dioxide from fossil fuels and cement manufacture in the year 2050?

d) What assumptions do we make when we construct a median-median line and use it to predict the values of a variable at some future date? For what reasons might such assumptions not be valid?

DRAWING INFERENCES

The median-median lines in the *Worked Example* suggest that the winning times for females and males in the 100-m dash will be about the same around the year 2061. Do you believe this prediction? Give reasons for your answer.

In the *Worked Example*, suppose we had used only the data prior to the 1996 Olympics. How would this change the intersection of the two median-median lines and the prediction?

Use your median line to predict the winning time for women in the 100-m dash in the year 2150. Do you believe this prediction? Explain your answer.

Copyright © 1998 by Brendan Kelly Publishing Inc.

EXPLORATION 7 Regression and the Line of Best Fit

THE BETTMANN ARCHIVE

Karl Friedrich Gauss 1777-1855

Karl Friedrich Gauss is regarded by historians of mathematics to have been a mathematician without peer (with the possible exceptions of Archimedes and Newton). The list of branches of mathematics that he developed or extended is too large for inclusion here. Suffice it to say that his genius for mathematics manifested itself when he was three years old and remained with him until his death in his 78th year.

When Gauss was 18 years old, he was developing a theory of errors in measurement. Out of this work, the ubiquitous *normal distribution* emerged (see Exploration 13). In exploring the question as to which of a large number of measurements is most likely the correct one, Gauss invented the *method of least squares.*

We learned in *Exploration* 6 that when two variables are strongly correlated, the points in their scatterplot cluster along a straight line. Of all possible straight lines, which one fits the data the best? To find the equation of this optimum line that we call the *line of best fit*, Gauss computed the values of the constants "a" and "b" in the equation

$$y = a + bx.$$

such that the sum of the squares of the vertical distances of the data points from the line with equation $y = a + bx$ is a minimum. (See graph on p 29.) The vertical distance of each data point (x_i, y_i) from the line is called a *residual*. Hence the line of best fit is merely the straight line for which the sum of the squares of the residuals is a minimum.

The importance of the line of best fit is that it enables us to predict the value of the y variable when we know or can measure the value of the x variable, and conversely (provided the two variables are strongly correlated). In 1855, Sir Francis Galton, investigating inherited traits, constructed a scatterplot of the heights of 928 adults against those of their parents. He discovered that these variables are strongly correlated. Furthermore, children of tall (short) parents were taller (shorter) than average but closer to the average than their parents. That is, extremes in height "regressed" closer to the mean in successive generations. Hence the line of best fit relating these two variables was called the *regression line.*

Often the variables x and y are related in a non-linear way. In such a case, we seek a regression curve that is not a straight line. For example, the quadratic regression curve is the parabola of best fit; that is the parabola with equation $y = ax^2 + bx + c$, where a, b, and c are chosen so that the sum of the squares of the residuals is a minimum. The equations (which define the regression curves) that you can access on your TI-83 are listed on your STAT CALC menu as shown here.

Linear regression curve	equation:	$y = ax + b$
Quadratic regression curve	equation:	$y = ax^2 + bx + c$
Cubic regression curve	equation:	$y = ax^3 + bx^2 + cx + d$
Quartic regression curve	equation:	$y = ax^4 + bx^3 + cx^2 + dx + e$
Linear regression curve	equation:	$y = a + bx$
Logarithmic regression curve	equation:	$y = a + b\ln x$
Exponential regression curve	equation:	$y = ab^x$
Power regression curve	equation:	$y = ax^b$
Logistic regression curve	equation:	$y = c / (1 + ae^{-bx})$
Sine regression curve	equation:	$y = a\sin(bx + c) + d$

Copyright © 1998 by Brendan Kelly Publishing Inc.

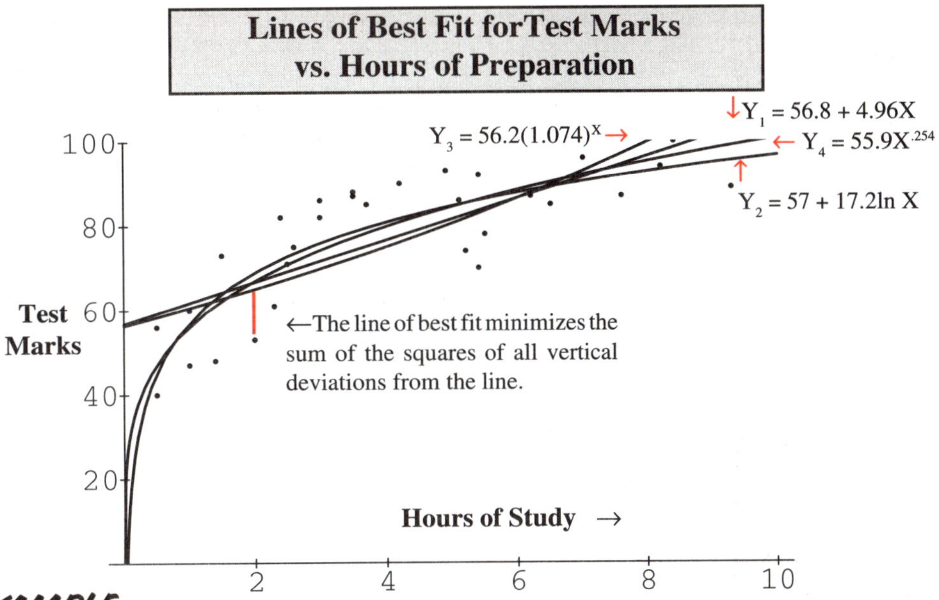

WORKED EXAMPLE

The graph above shows a scatterplot of the data on page 22, relating the marks on a mathematics test with the number of hours of study preparation.

a) Use the data on page 22 to create a scatterplot like the one above on your TI-83.
b) Find the equations of the linear and the logarithmic regression lines.
c) Graph the regression lines so they are superimposed on the scatterplot.
d) Which of the regression curves yields the better fit as defined by Gauss?
Use that line to estimate the mark a student would achieve with 3.2 hours of study.

SOLUTION

a) We follow the procedure in *Worked Example* 1 on page 22 to enter the ordered pairs into the STAT EDIT table and to graph the scatterplot.

b) To calculate the equation of the regression line (with equation of the form Y = a + bX) we press: [STAT] [▶] [4] [ENTER] . We obtain equation of the regression line shown in the display. We define this equation as Y_1 on the [Y =] list by pressing:

[Y =] [VARS] [5] [▶] [▶] [ENTER]

We proceed as above to define the expression for the logarithmic regression curve as Y_2. That is, we press: [STAT] [▶] [9] [ENTER] and proceed as above.

c) When we press: [ZOOM] [9] we get the display shown here.

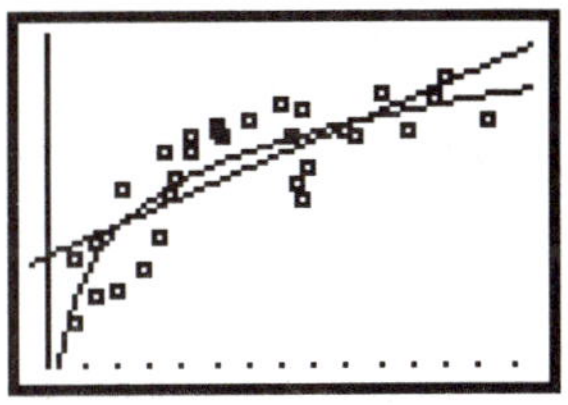

d) To determine which curve is the better fit for the data, we calculate the residuals; that is the difference between the observed value (given data) and the predicted value (y coordinate on the regression line). To calculate the list of residuals L_3 for the linear regression, we proceed as above to obtain the linear regression equation. The residuals are automatically calculated and stored in memory under the name **RESID**. To store these residuals to list L_3, we enter:

[2nd] [LIST]. We select RESID and press: [ENTER] [STO▶] [2nd] [L_3] [ENTER]

We evaluate the sum of the squares of the residuals by evaluating **1-Var Stats** with the Xlist as L_3. This yields the display shown here.

Proceeding as above, we store the residuals of the logarithmic regression equation to L_4 and obtain the sum of squares of the residuals of residuals L_4 shown in the display. Since the sum of squares of the residuals is less for the logarithmic curve, it provides the better fit. We then store 3.2 in variable X and evaluate Y_2 to get an estimated mark of about 77.

Copyright © 1998 by Brendan Kelly Publishing Inc.

Exercises

1. Write a sentence or two to explain the meaning of each of the following terms.

- method of least squares
- regression line
- regression curve
- line of best fit
- residual

2. For each of the following scatterplots estimate the slope of the straight line of best fit. (Assume the x and y-scales are identical.)

3. Explain how a median-median line and a regression line are alike and how they are different.

4. Write the general equations of the curves of best fit (using "a" and "b" as parameters) in each of the following: (The first one is done for you.)

- the linear regression model $Y = a + bX$
- the logarithmic regression model
- the exponential regression model
- the power regression model

5. Explain how you can determine which regression curve of several different types fits a given set of data best.

6. In the worked example, we compared the regression line and the logarithmic regression curve in fitting the data pertaining to hours of study and test marks.

a) Would you expect the relationship between hours of study and test marks to have been linear? Give reasons for your answer.

b) Why might you expect the relationship between hours of study and exam mark to be logarithmic?

7. a) Using the table on page 25, enter the Olympic year in column L_1 of your STAT EDIT table and the men's winning times in column L_2. Then graph a scatterplot.

b) Calculate the linear and logarithmic regression equations. Which regression curve, the linear or the logarithmic, provides a better match to the data? Why?

c) Use the regression curve of better fit to estimate the men's winning time at the Summer Olympics in 2060. Compare this estimate with the median-median line estimate on page 26. Explain reasons for any difference in the two estimates.

Investigations

8. Using the data in Exercise 3 on page 24, construct a scatterplot for the ages of women and their husbands.

a) What is the correlation between the age of a woman and her spouse? Is this a strong correlation? Calculate and graph the equation of the regression line for this scatterplot.

b) Graph the logarithmic regression curve. Which curve, the linear or the logarithmic, provides a better fit to the data? Why do you think this curve provides a better fit?

c) Use the better regression curve to estimate the age of a man who is married to a 48-year-old woman.

9. The ordered pairs in the table below show the U.S. population at the beginning of each decade from 1800 to the present. (The first coordinate gives the year and the second coordinate the population in millions.)

(1800, 5.31)	(1810, 7.24)	(1820, 9.64)	(1830, 12.9)
(1840, 17.1)	(1850, 23.2)	(1860, 31.4)	(1870, 39.8)
(1880, 50.2)	(1890, 62.9)	(1900, 76.0)	(1910, 92.0)
(1920, 105.7)	(1930, 122.8)	(1940, 131.7)	(1950, 150.7)
(1960, 179.3)	(1970, 203.3)	(1980, 226.5)	(1990, 248.7)

Make a scatterplot of the U.S. population vs the date. Which regression curve yields the best fit to this data? Use it to estimate the U. S. population in 1965 and in 2010.

10. The table below shows the ages and the systolic blood pressures of 10 women.

Age vs. Systolic Blood Pressure for a Sample of 10 Women

Age	Blood Pressure	Age	Blood Pressure
34	120	55	140
38	114	57	152
44	140	60	157
44	136	63	149
49	147	72	162

Graph a regression curve that is a best fit of the data. Use your regression curve to estimate the blood pressure of someone who is 80 years of age.

DRAWING INFERENCES

Collect 20 measurements of any two variables that you believe may be correlated. Construct a scatterplot of your data. Record the correlation coefficient of the two variables.

If the two variables have correlation r with $|r| > 0.3$, construct a curve of best fit and estimate the dependent variable for a given value of the independent variable.

Copyright © 1998 by Brendan Kelly Publishing Inc.

UNIT 3

EXPERIMENTAL PROBABILITY & SIMULATIONS

In your first encounter with the study of chance, you learned how to calculate the probability of an event given the theoretical probabilities of the outcomes favorable to it. The probabilities of the outcomes were either given or easily calculated. However such is not always the case. Often we wish to know the probability of an event that cannot be described in terms of simple outcomes of known probability. For example, we may wish to know the probability that a radiation treatment will abort a tumor or the probability that a shuffled deck will yield a win in a game of solitaire. In such cases, we can conduct an experiment or run a simulation to gather data and calculate the relative frequency of a particular event on a large number of trials. When the number of trials is sufficiently large, we can assume that the relative frequency of that event is close to its "true probability", and we call this the *experimental probability* of that event. In this unit, we will present some simple programs that will enable you to determine experimental probabilities of events by simulating experiments and calculating relative frequencies.

To enter into your TI-83 the programs in this unit, you can either download them from the teacher's calculator using the LINK cable, or you can enter them yourself using the menus shown here.

To create a new program, press: PRGM ▶ ▶ ENTER. We obtain

```
PROGRAM
Name=
```

Upon typing a name and pressing ENTER, we obtain a blinking cursor, indicating that the calculator is ready to take programming instructions. The main programming commands are shown on the three menus below.

```
PROGRAM:ANYTHING
:
```

THE CONTROL COMMANDS

← To access, press PRGM when in program mode.

```
CTL I/O EXEC
1:If
2:Then
3:Else
4:For(
5:While
6:Repeat
7:End
8:Pause
9:Lbl
0:Goto
A:IS>(
B:DS<(
C:Menu(
D:prgm
E:Return
F:Stop
G:DelVar
H:GraphStyle(
```

For(J, 1, 13, 2) means: "Let **J** take the values 1, 3, 5, 7, 9, 11, 13 in turn, and execute the commands between the **For** and **End** statements until **J** reaches 13. Then proceed to the statement following the **End** statement". The **For(... End** sequence of commands is called a *loop*.

THE INPUT/OUTPUT COMMANDS

← To access, press PRGM ▶ when in program mode.

```
CTL I/O EXEC
1:Input
2:Prompt
3:Disp
4:DispGraph
5:DispTable
6:Output(
7:getKey
8:ClrHome
9:ClrTable
0:GetCalc(
A:Get(
B:Send(
```

To prompt the program user to enter a number in response to a question, we use the **Input** command followed by the question in quotes and the letter to which we want the answer stored. To display the value of a variable Z on the screen, we use the command **Disp Z**.

THE RANDOM NUMBER COMMANDS

← To access, press MATH ◀

```
MATH NUM CPX PRB
1:rand
2:nPr
3:nCr
4:!
5:randInt(
6:randNorm(
7:randBin(
```

rand(3) generates 3 random numbers between 0 and 1

randInt(2,5,3) generates 3 random integers between 2 and 5.

Copyright © 1998 by Brendan Kelly Publishing Inc.

EXPLORATION 8 Experimental Probability

Theoretical probability is based on the idea that certain outcomes are equally likely to occur. That is, if an experiment has n equally likely possible outcomes, we can assign a probability of $\frac{1}{n}$ to each outcome. The theoretical probability of more complex outcomes or *events* can then be expressed in terms of the probabilities of these simple outcomes. However, in some experiments, such as the tossing of a paper cup (see exercise 2). The probabilities of the possible outcomes are not known and must be estimated by experiment. To provide a basis for the concept of *experimental probability*, we define *relative frequency*.

"No one said anything about life being fair."

> Definition: If an outcome A occurs f times on n trials of an experiment, then:
> the *relative frequency* of outcome A = $\frac{f}{n}$

The cartoon suggests that the concept of "fair" may be unrealizable in the real world. In the *Worked Examples*, we investigate the relative frequencies of the outcomes in the toss of a fair coin as the number of tosses is increased.

Using the concept of relative frequency, we can define the *experimental probability* of an outcome as follows:

> Definition: The *experimental probability* of an outcome A is the relative frequency of A for a large number of trials.
> In more formal notation, we would write:
>
> $$\textit{experimental probability of } A = \lim_{n\to\infty} \frac{f}{n}$$
>
> (This definition is not rigorous, since a formal limit does not exist.)

A *fair* coin is defined to be a coin for which the heads and tails outcomes are equally likely. Time constraints make it impractical to conduct experiments involving a large number of coin tosses. To *simulate* the tossing of a fair coin, we must replace it with an experiment having two equally likely outcomes. The program shown below for your TI-83 graphics calculator simulates the tossing of a fair coin by selecting an integer randomly from the interval, [0, 1]. Since 0 and 1 are equally likely, we can associate the outcome 0 with tails and the outcome 1 with heads. This program asks for the number of trials (tosses) requested, runs these trials and then reports the number of heads obtained.* (Alternatively, we could use the command **randBin(** to simulate the tossing of a coin for N trials. This command is described in your TI-83 manual.)

To create program **COINTOSS**, press: PRGM ◄ ENTER and key in : COINTOSS one letter at a time and press ENTER.

To access **Input**, press: PRGM ► ENTER

To access **For**, press: PRGM 4 ENTER

To access **Disp**, press: PRGM ► 3

```
PROGRAM:COINTOSS
:Input "HOW MANY
 TOSSES?",N
:0→H
:For(J,1,N,1)
:randInt(0,1)→X
:H+X→H
:Disp H
:End
:Disp "NUMBER OF
 HEADS ",H
:
```

To space between words, press ALPHA [␣].

To access the arrow, press STO►

The **randInt(** command is accessed by pressing: MATH ◄ 5

*For more instruction on how to enter and create programs, see our publication: *Exploring Advanced Algebra with the TI-83.*

Copyright © 1998 by Brendan Kelly Publishing Inc.

WORKED EXAMPLE 1

Enter into your TI-83 calculator the program **COINTOSS**.

a) Run **COINTOSS** for n tosses of a fair coin when n is equal to: 50, 100, 200, 500 and 1000. Create a table to display n and the number of heads for each n.

b) Calculate the relative frequency of heads in each case and display in your table.

c) Make a scatterplot of the number of heads vs n.

d) Determine the experimental probability of "heads" from your data.

SOLUTION

To enter the program, we proceed as described on page 32 and press 2nd [QUIT] to exit the programming mode.

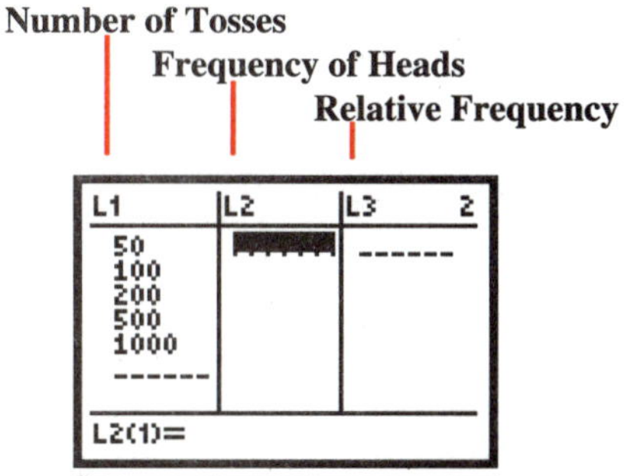

a) To create a 3-column table with headings, L_1, L_2, and L_3 where L_1 is the number of tosses, L_2 the number of heads and L_3 the relative frequency of heads, press STAT ENTER . Clear columns labeled L_1, L_2, and L_3. Then in column L_1 enter the list 50, 100, 200, 500, and 1000 as shown in the display.

To run **COINTOSS** for 50 tosses, press: PRGM; move the cursor opposite **COINTOSS** and press ENTER ENTER .

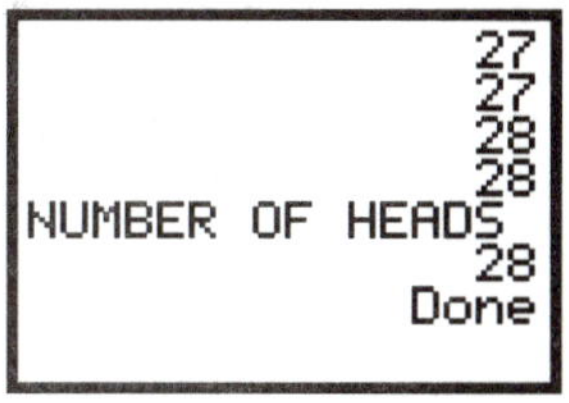

In response to the prompt **HOW MANY TOSSES?** we enter **50** ENTER . We obtained the display on the right showing that 28 of the tosses yielded heads. To enter 28 in the list L_2, we press: STAT ENTER and use the cursor keys to position the cursor in column L_2 to enter 28. To run COINTOSS for n = 100, we press 2nd [QUIT] ENTER .

Repeating this procedure for n = 100, 200, 500 and 1000, we obtained the table in the display.

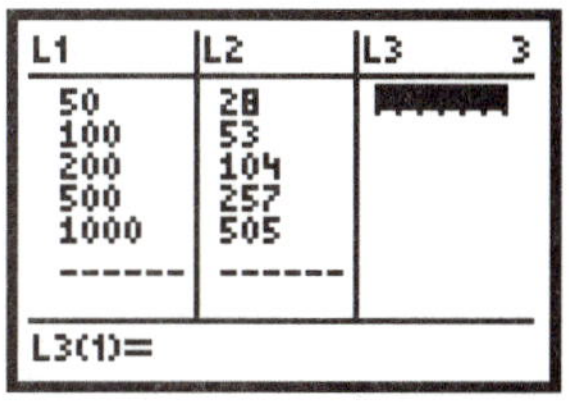

b) To display the relative frequencies in column L_3, we quit the STAT mode and press:

2nd [L_2] ÷ 2nd [L_1] STO► 2nd [L_3] ENTER

To return to the table, we press: STAT ENTER and obtain the display on the right.

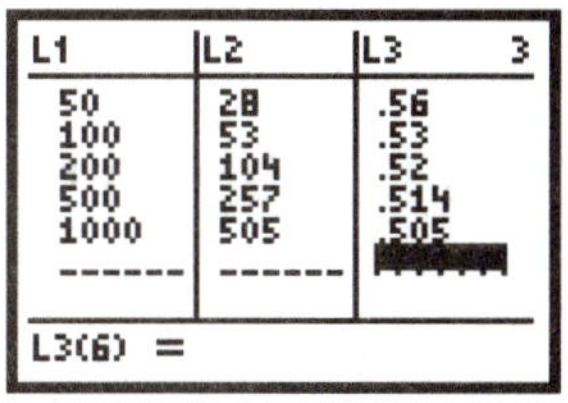

We observe that the relative frequencies seem to approach 0.5 as the number of tosses, n, increases.

c) We set the window variables to $0 \le x \le 1000$; $0 \le y \le 1000$ and press: 2nd [STAT PLOT]
We define a plot by selecting the scatterplot option and defining the **Xlist** and **Ylist** as L_1 and L_2 respectively.

Upon pressing GRAPH we obtain the display on the right.
The points in the scatterplot cluster along a straight line of slope 0.5.

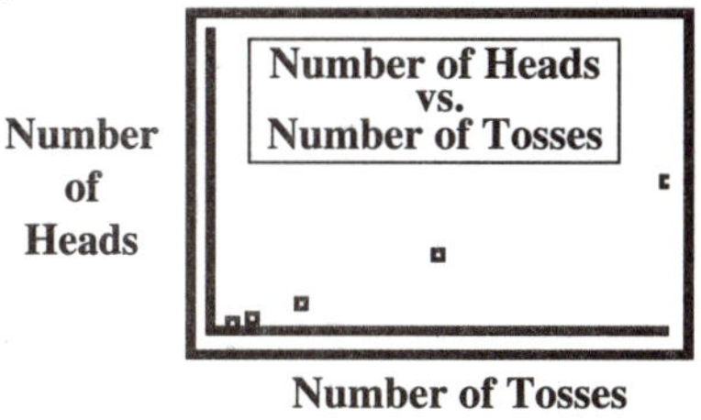

d) To obtain the experimental probability of the outcome, HEADS, we plot the relative frequency of HEADS as the number of tosses, n, increases. The window is changed to $0 \le x \le 1000$; $0 \le y \le 1$. We obtain the display on the right, which shows visually what we saw in the third column of the table; i. e. that the relative frequency of HEADS seems to approach 0.5 as the number of tosses, n, increases.

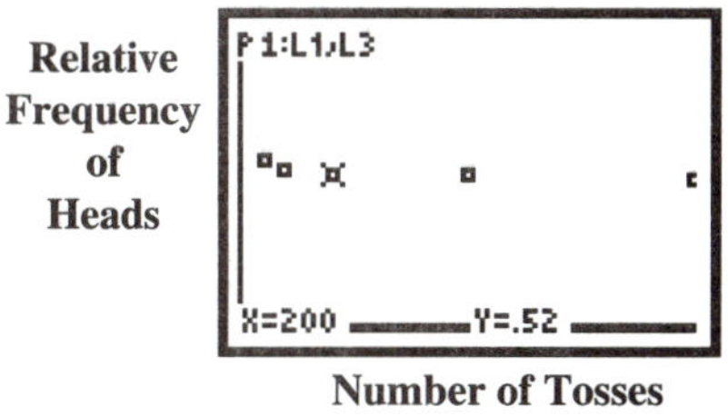

In all we simulated the toss of a coin 1850 times and observed 947 heads, so the experimental probability of heads obtained in this simulation is 947/1850 or 0.51… Of course, different replications of this simulation will yield different results, although most of the time the relative frequency will be close to 0.5.

Copyright © 1998 by Brendan Kelly Publishing Inc.

BETTMANN

Albert Einstein
1879-1955

Albert Einstein, the creator of the Special and General Theories of Relativity, revolutionized modern physics. His discovery of the equivalence of mass and energy in 1905 captured in his famous equation $E = mc^2$ paved the way for the generation of nuclear energy — a discovery that promises significant benefits to humanity as fossil fuels are depleted. His re-formulation of Newton's laws of motion and his expression of gravitational fields as mathematical tensors resolved some unanswered questions that had confounded classical physics. Although revolutionary in his approach to physics, Einstein never accepted the interpretations of the new Quantum Physics that called into question strict causality and introduced a dimension of chance into physics. His skepticism is evident in his oft quoted statement,

I shall never believe that God plays dice with the world.

The role of chance in physics and in life itself has fascinated humans since the beginning of history. Dice carved from bones or the horns of animals have been used in games of chance. The six faces of the cube have been considered equally likely outcomes when the cube is rolled and consequently dice are used in most modern board games. When a large number of rolls of a die are required, it is useful to simulate the roll of a die with the random draw of a number from the set {1, 2, 3, 4, 5, 6}. The program DIETOSS asks for input regarding N, the number of simulations of the toss of a die that you wish to run. It draws an integer at random from the interval [1, 6] using the **randInt(** command on the PRB (probability) submenu of the MATH menu of your TI-83. It then displays a histogram showing the frequency of each outcome on the N "tosses".

WORKED EXAMPLE 2

Enter into your TI-83 calculator the program **DIETOSS**.
- a) Run DIETOSS to simulate 100 tosses of a die.
- b) Trace along the histogram to determine the frequency of the outcome which occurs most often.
- c) Display a table showing the frequency of each outcome.

SOLUTION

To create program **DIETOSS**, we press: PRGM ◄ ENTER and key in: DIETOSS one letter at a time and press ENTER.

```
PROGRAM:DIETOSS
:Input "HOW MANY
 TOSSES?",N
:1→Xmin:7→Xmax
:0→Ymin
:Fix 0
:{1,2,3,4,5,6}→L
1
:0L1→L2
:For(I,1,N)
:randInt(1,6)→X
:L2(X)+1→L2(X)
:End
:max(L2)+1→Ymax
:Plot1(Histogram
,L1,L2)
:FnOff
:1→Xscl
:DispGraph
:
```

This sets the mode to 0 decimal digits. To access this press: MODE

Press: STO► 2nd [L_2]

To access **For**, press: PRGM 4 ENTER

To access **End**, press: PRGM 7 ENTER

The **randInt(** command is accessed by pressing: MATH ◄ 5

2nd [STAT PLOT] ENTER
2nd [STAT PLOT] ► 3

1 STO► VARS ENTER 3 ENTER

PRGM ► 4 ENTER

a) We run DIETOSS and enter 100 in response to the prompt. In the case shown here, we obtain the histogram below left.

b) We press: TRACE and move the cursor to the tallest bar. The display shows that the frequency of outcome 5 was 22.

c) When we press: STAT ENTER we get the table below.

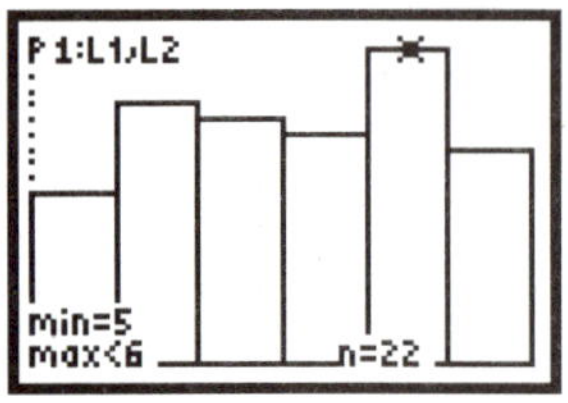

L1	L2	L3 2
1	12	1
2	18	1
3	17	1
4	16	1
5	22	1
6	15	

L2(5) =22

Copyright © 1998 by Brendan Kelly Publishing Inc.

Exercises

1. Explain the difference between experimental probability and theoretical probability.

2. When a paper cup is tossed, there are 3 possible outcomes. The cup can land…

a) upright b) upside down c) on its side

Explain how you would obtain the experimental probability of each outcome.

3. The probability that a randomly chosen American citizen will ultimately die from heart disease is about 0.43.

a) How is this probability determined?

b) Is this a theoretical probability or an experimental probability? Explain why this probability is used.

Enter the program COINTOSS, given on page 32, by following the procedures given in *Worked Example* 1. Use this program to help you complete Exercises 4 and 5.

4. a) Run your program to simulate 100 tosses of a fair coin. Record the frequency of heads. Repeat this experiment 9 more times and record the frequency of heads in each case.

b) Find the median frequency of heads for the 10 simulations. Explain why one might expect to find a median frequency close to 50.

5. a) Run your program to simulate 50, 100, 200, 500, and 1000 tosses of a fair coin. Display the number of tosses, frequencies, and relative frequencies in a table as in *Worked Example* 1.

b) Compare the relative frequencies of heads in your table with those in *Worked Example* 1 for 50 tosses and for 1000 tosses. Are the relative frequencies closer for $n = 50$ or for $n = 1000$? Is this what you would expect? Explain.

6. a) In a game called the *Problem of the Points*, Players A and B each toss a coin. If either coin shows heads, then A wins. Otherwise B wins. What is the probability that A wins? The program TWOCOINS simulates N trials of this game. Enter this program into your calculator and run it for 100 games. Record the number of times that A wins.

```
PROGRAM:TWOCOINS
:Input "HOW MANY
 GAMES? ",N
:0→W
:For(I,1,N,1)
:randInt(0,1,2)→
L₁
:If L₁(1)+L₁(2)>
0
:W+1→W
:End
:Disp "WINS FOR
A ",W
:
```

b) Repeat the procedure in a) nine more times and record the relative frequency of wins for A.

c) Roberval asserted that there are only three possible outcomes: two heads, two tails and one of each. Hence A would win 2/3 of the time. Does your experiment support Roberval's assertion?

Investigations

7. a) If H and T denote respectively the frequencies of heads and tails in the toss of a fair coin, estimate the value of $|H - T|$ after 100, 500, and 1000 tosses of a fair coin.

b) Use your COINTOSS program to determine whether your estimate is reasonable.

c) Does $|H - T|$ approach 0 as the number of tosses becomes very large? Explain your answer.

8. Enter the program DIETOSS into your calculator. (See *Worked Example* 2.)

a) Run DIETOSS to simulate 100 tosses of a die. Trace along the histogram to determine the frequency of each outcome.

b) Display the frequency and the relative frequency of each outcome in a table.

c) Compare the relative frequencies you obtained with those presented in *Worked Example* 2. Compare the experimental probability of each outcome with its theoretical probability.

d) Repeat the simulation for 1000 tosses and display in a table, the relative frequencies of all the outcomes. What do you discover about the relative frequencies of the outcomes as the number of tosses increases? (Retain COINTOSS, TWOCOINS and DIETOSS in for use in *Exploration* 9.)

9. Calculate the theoretical probability of each event on the roll of a die.

a) an even number occurs

b) a number greater than 4 occurs

c) a 5 does not occur

d) either a 3 or a 6 occurs

Run DIETOSS to simulate 1000 rolls of a die to determine the experimental probability of each of the above events. Compare the theoretical probability and the experimental probability of each event. Compare the theoretical and experimental probabilities for a simulation of 5000 rolls of a die.

DRAWING INFERENCES

Use the COINTOSS simulation to estimate the probability that a family of 7 children has exactly 4 boys. (Assume that both sexes are equally likely at birth.)

Copyright © 1998 by Brendan Kelly Publishing Inc.

EXPLORATION 9 *Simulations & The Random Walk*

In 1827 a botanist, Robert Brown, observed through his microscope the erratic motions of pollen dust in water. He hypothesized that the dust particles were under bombardment from the water molecules. This effect has later become known as *Brownian motion.* Since the molecular theory of matter was considered highly speculative, Brownian motion received little attention until the turn of the century.

© 1997 by Sidney Harris

In 1905, Albert Einstein published four epoch making papers in *Annalen der Physik.* One of those papers presented a mathematical model for Brownian motion. The subsequent extensions of Einstein's work by other mathematicians such as Norbert Wiener produced mathematical models that provided accurate representations of diffusion processes. Today diffusion processes are modeled by simulations such as the *random walk.*

In 1921, Albert Einstein won the Nobel Prize, not for the mass-energy equivalence discovery, nor for his discovery of the Theory of Relativity, nor for his work on Brownian motion.
For what work did Einstein receive his Nobel prize?[1] (See footnote.)

In the one-dimensional random walk, a particle bombarded by a molecule is moved to either the right or the left with equal probability. The question to be investigated is, *What is the long term behavior of the particle when the number of bombardments becomes very large?* Will the particle end up close to where it began, will it eventually move away from its initial point never to return, or will it oscillate back and forth between these extremes? To simulate this problem, we consider a particle which is situated at 0 on a number line. We then toss a coin and move 1 unit right if the coin shows heads and 1 unit left if the coin shows tails. Where do you think the particle will be after 100 tosses of the coin? Where do you think it will be after 1000 tosses?

Where Will the Particle Be After 5000 Tosses of the Coin?

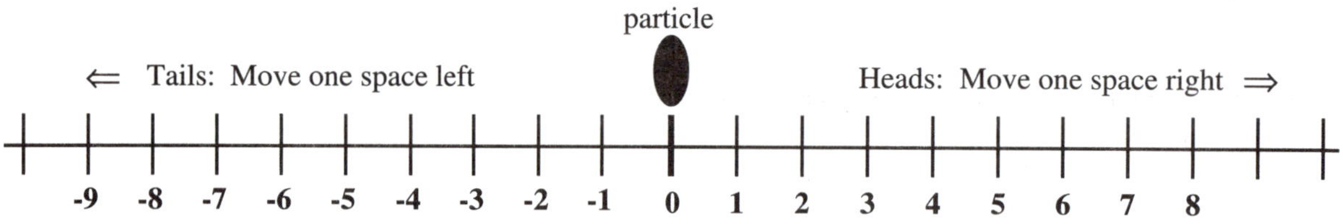

WORKED EXAMPLE 1

Run COINTOSS to simulate 100, 1000, and 5000 bombardments of the particle.
Display the final positions of the particle in a table.
Describe how the probability that the particle will finish near 0 changes as the number of tosses becomes vey large.

SOLUTION

When we run COINTOSS with N = 100, 1000, and 5000 respectively, we obtain the frequencies of heads shown in the table in column L_2. The corresponding final positions of the particle are shown in column L_3 where $L_3 = 2L_2 - L_1$, which gives the number of heads minus the number of tails ($L_2 - [L_1 - L_2] = 2L_2 - L_1$).

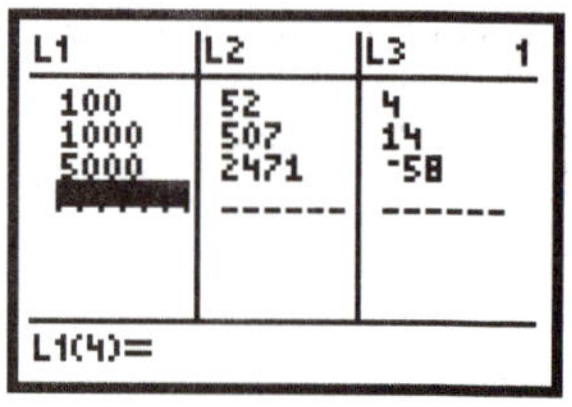

The numbers in column L_3 suggest that the final distance from 0 increases with N. As N increases, there is an increased number of final positions which the particle can occupy, so the probability that it will occupy a final position close to 0 decreases.

[1] Einstein's 1905 paper that gave a mathematical model explaining the photoelectric effect won him the Nobel Prize in 1921.

Copyright © 1998 by Brendan Kelly Publishing Inc.

WORKED EXAMPLE 2

Run the simulation COINTOSS to estimate the probability that there will be exactly 3 girls in a family of 5 children. Compare this estimate with the theoretical probability.

SOLUTION

Since both sexes are equally likely at birth, we can simulate the selection of a sex by the toss of a coin. Getting exactly 3 girls out of 5 children has the same probability as achieving 3 heads out of 5 tosses of a fair coin. Therefore, to simulate the assignment of sexes to 5 children, we run COINTOSS for 5 tosses. To do this, press: **PRGM** **1** **ENTER**

(**1** — or whatever the number of program **COINTOSS** in the program list.)

In response to the prompt, HOW MANY TOSSES?, press: **5** **ENTER**
Almost immediately the calculator displays: NUMBER OF HEADS followed by a number between 0 and 5 on the line below. This represents the number of girls in that selection of 5 children. We run this experiment 99 more times, by pressing **ENTER** **5** **ENTER** each time and forming a tally like the one on the right.
The simulation gives an experimental probability of about 0.38. The theoretical probability is 0.3125. By increasing the number of trials, we would probably approximate the theoretical probability even more closely.

Number of Trials	Freq of outcome 3	Relative Frequency
卌 卌 卌 卌 卌 卌 卌 卌 卌 卌 卌 卌 卌 卌 卌 卌 卌 卌 卌 卌	卌 卌 卌 卌 卌 卌 卌 ///	38 out of 100 = 0.38

The program BINOMIAL in *Worked Example* 3, automates the tedious tally process in *Worked Example* 2. BINOMIAL displays in a histogram the frequency of each outcome in the tossing of N coins repeated M times.

WORKED EXAMPLE 3

Enter program BINOMIAL into your calculator. Run BINOMIAL to estimate the probability that there will be exactly 3 girls in a family of 5 children. Compare your estimate with the estimate in *Worked Example* 2 and with the theoretical probability.

SOLUTION

We enter program BINOMIAL using the menus described in the Worked Examples of this and the previous exploration. When we run BINOMIAL, we enter 5 in response to the prompt HOW MANY COINS?. We enter 100 in response to HOW MANY TOSSES?.

Within 30 seconds, the display shows a histogram like that below. There are 6 bars representing the frequencies of the 6 possible outcomes: 0 heads, 1 head, 2 heads,…5 heads. (The first bar is obscured by the text on the display.) We press **TRACE** and move the cursor to the fourth bar which represents the frequency of exactly 3 heads. The display shows n = 29, so the experimental probability is 0.29 compared to 0.3125, theoretical probability, and 0.38 in *Worked Example* 2. Press **STAT** **ENTER** to display the table of outcomes (L_1), frequencies (L_2), and relative frequencies (L_3).

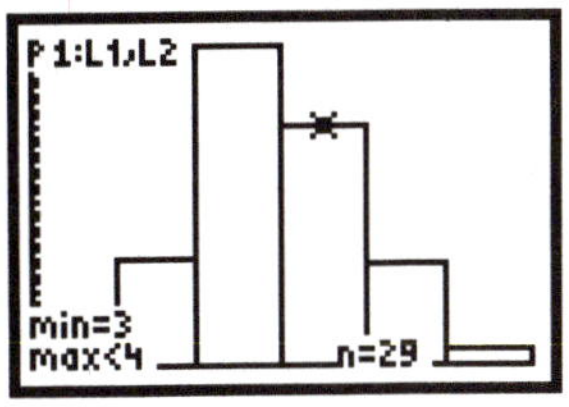

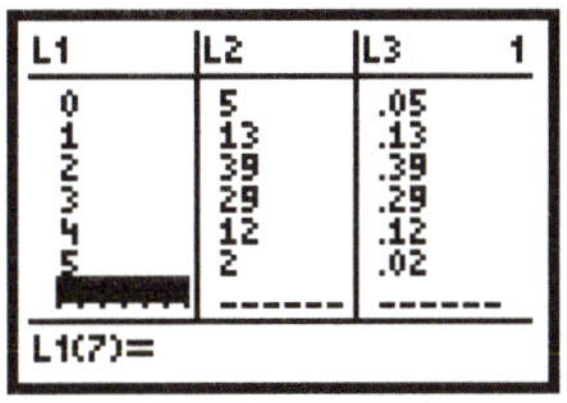


```
PROGRAM:BINOMIAL
:Float
:Input "HOW MANY
 COINS?",N
:Input "HOW MANY
 TOSSES?",M
:0→Xmin:N+1→Xmax
:1→Xscl
:seq(X,X,0,N,1)→
L1
:0L1→L2
:For(I,1,M)
:randBin(N,.5)→X
:L2(X+1)+1→L2(X+
1)
:End
:max(L2)+1→Ymax
:L2/M→L3
:PlotsOn 1
:max(L2)+1→Ymax
:L2/M→L3
:PlotsOn 1
:Plot1(Histogram
,L1,L2)
:DispGraph
:
```

Copyright © 1998 by Brendan Kelly Publishing Inc.

The previous examples show how we can use a coin toss to simulate processes involving a number of repetitions of an event that has two equally likely outcomes. The next example shows how we can use a simulation to estimate the area of an ellipse. To perform this simulation, we use program **ELLIPSE** which selects points randomly from the rectangle defined by: $-A \le x \le A$; $-B \le y \le B$ and records the proportion of those points which lie inside or on the ellipse with equation: $B^2x^2 + A^2y^2 = A^2B^2$

WORKED EXAMPLE 4

Run program ELLIPSE to estimate the areas of the ellipses with these equations:

a) $9x^2+16y^2=144$ b) $x^2+4y^2=4$ c) $9x^2+y^2=9$

Use your answers to conjecture a formula for the area of an ellipse with equation: $B^2x^2 + A^2y^2 = A^2B^2$

SOLUTION

```
PROGRAM:ELLIPSE
:Fix 3
:PlotsOff
:Prompt A,B
: -A-1→Xmin
:A+1→Xmax
: -B-1→Ymin
:B+1→Ymax
:Input "HOW MANY
 POINTS? ",N
:0→S
:"(B/A)√((A²-X²)
"→Y1
:"(-B/A)√((A²-X²
)"→Y2
:DispGraph
:Line(-A,-B,-A,B
)
:Line(-A,B,A,B)
:Line(A,B,A,-B)
:Line(A,-B,-A,-B
)
:For(I,1,N,1)
:2Arand-A→X
:2Brand-B→Y
:Pt-On(X,Y)
:If B²X²+A²Y²<A²
B²
:S+1→S
:End
:S/N→T
:Pause
:Disp "PROPORTIO
N"
:Disp "OF PTS IN
SIDE",T
```

a) When we run **ELLIPSE**, we enter 4 and 3 respectively in response to the prompts **A?** and **B?**, because these are the values of the x and y-intercepts. We enter 500 in response to the prompt, **HOW MANY POINTS?**. This causes **ELLIPSE** to select 500 random points from the rectangle, $-A\le x\le A$; $-B\le y\le B$ and test each one to determine whether it is inside the ellipse. As the points are selected, they are displayed on the screen as shown in the display below.

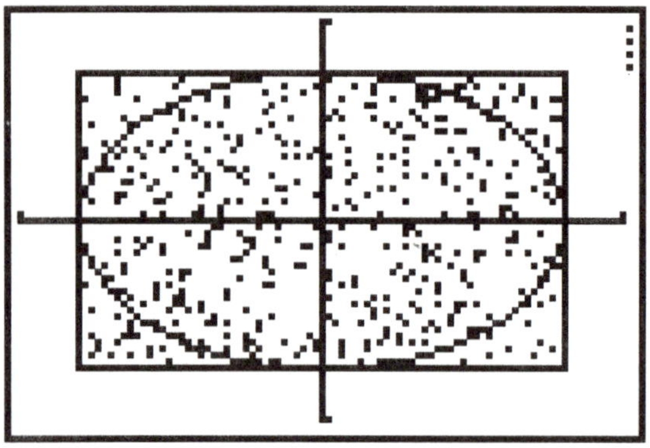

←When the cursor in the upper right corner changes from the busy to the pause mode, we press ENTER and we obtain the display shown below.

```
B=?3
HOW MANY POINTS?
500
PROPORTION
OF PTS INSIDE
            .772
            Done
```

This display indicates that of all the points randomly selected, 77.2% of them were inside the ellipse. Since all points are equally likely, it would seem that the area of the ellipse must be approximately 77.2% of the area of the bounding rectangle. The area of the bounding rectangle is 4**AB** or 4(4)(3) = 48 square units, so the area of the ellipse is about 0.772×48 or 37.1 square units. (The actual area of the ellipse is 37.7 square units.)

b) We repeat the procedure in part a) with **A** = 2; **B**=1 and **N**= 300. We obtain the displays from which we obtain an estimated area of 0.760×8 or 6.1 square units.

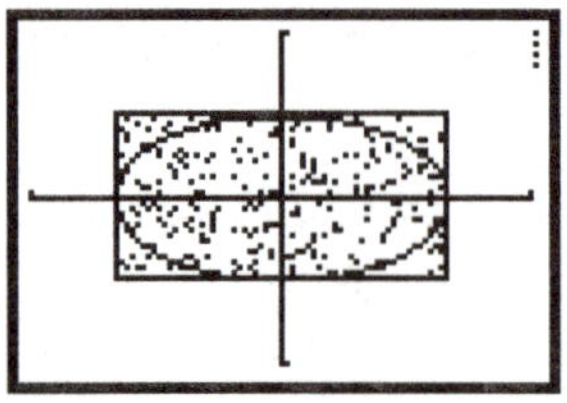

```
B=?1
HOW MANY POINTS?
300
PROPORTION
OF PTS INSIDE
            .760
            Done
```

c) We repeat the procedure in part a) with **A** = 1; **B**=3 and **N**= 300. We obtain the displays from which we obtain an estimated area of 0.803×12 or 9.6 square units.

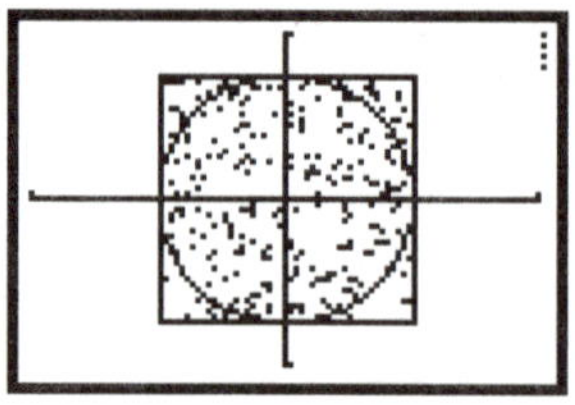

```
B=?3
HOW MANY POINTS?
300
PROPORTION
OF PTS INSIDE
            .803
            Done
```

When we run **ELLIPSE** for large numbers of points, we find that the proportion approaches 78.5%. The estimate of the area of the ellipse is therefore $0.785(4AB) \approx 3.14AB$. Look familiar?

Copyright © 1998 by Brendan Kelly Publishing Inc.

Exercises

1. What is a simulation? Explain how simulations are used to obtain experimental probabilities.

2. A simulation is based on the idea that the experimental probability of each outcome will be approximately equal to its theoretical probability (if one exists). However, the laws of chance dictate that occasionally there will be a significant gap between an experimental probability and the corresponding theoretical probability. Is there any way to be sure that such a discrepancy does not occur? Explain.

Use **COINTOSS, DIETOSS** or the **BINOMIAL** simulation to help you complete Exercises 3 through 8.*

3. a) Estimate the probability of guessing exactly 7 correct answers on a 10-item true-false test.
b) Estimate the probability of guessing at least 7 correct answers on a 10-item true-false test.

4. a) Estimate the probability that there are exactly 4 girls in a family of 6 children.
b) Estimate the probability that there are at least 4 girls in a family of 6 children.

5. Each item on a 20-item multiple choice test offers 6 possible answers. Estimate the probability that a student will guess correctly on more than one quarter of the items.

6. In a particular occupation, one third of all employees are female. The Acme Company has 82 employees in that occupation but only 21 of them are female. Estimate the probability that this gender balance could have occurred by chance.

7. An airline books 96 passengers on its jet liner which has capacity for only 85 people. If there is only one chance in 6 that a passenger will not show for the flight, what is the probability that more than 85 of the 96 passengers will show?

8. The members of the Old Timers Hockey League draw slips of paper before each game to determine which of the 6 positions each will play. Estimate the probability that a particular member will have played all 6 positions after 20 games.

9. a) Run **ELLIPSE** to estimate the area of the ellipses defined by these equations.

i) $5x^2 + 8y^2 = 40$ ii) $3x^2 + 9y^2 = 27$ iii) $12x^2 + 7y^2 = 97$

b) Compare you estimates with the actual area given by:

$$\text{Area} = \pi AB$$

where A and B are the x and y-intercepts respectively.
Explain how your estimate changes as the number of points selected is increased.

Investigations

10. a) Run **ELLIPSE** with $A = \sqrt{2}$, $B = \sqrt{7}$ and $N = 1000$. Use the output to estimate the area of the ellipse defined by the equation: $7x^2 + 2y^2 = 14$.

b) Modify **ELLIPSE** to create program **HCYCLOID,** which estimates the area inside the hypocycloid defined by the equation:

$$x^{2/3} + y^{2/3} = c \quad \text{where } c \text{ is a constant}$$

c) Run **HCYCLOID** to estimate the area in the first quadrant enclosed by the hypocycloid with $c = 3$.

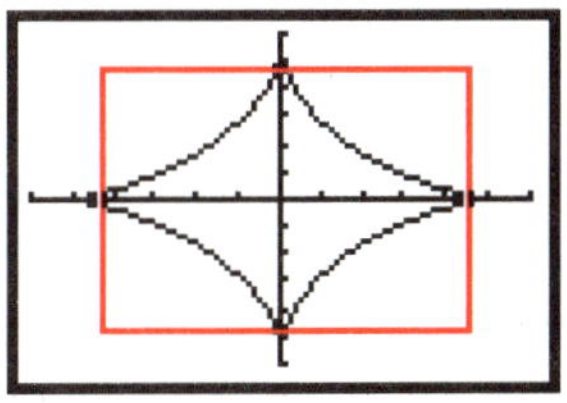

d) The formula for the area in the first quadrant enclosed by a hypocycloid is:
Area = $k\pi c^3$ where c is the constant in the equation.
Use **HCYCLOID** to estimate k.

11. a) Run **BINOMIAL** for N = 10000 to estimate the probability that there will be exactly 4 boys in a family of 7 children. (This will take a couple of hours to run, so do it before you go to bed and let the program run while you sleep. You can harvest the results in the morning.)

b) Calculate the theoretical probability and compare with your experimental probability.

12. a) Modify **DIETOSS** to create **MTOSS,** a program that creates the same kind of histogram as **DIETOSS,** but for an M-sided die.

b) An employer discovers that 496 of the 2017 absences reported last month occurred on Fridays. Run **MTOSS** to estimate the probability that this would occur from illness alone, assuming that illness occurs on any of the five work days of the week with equal probability.

CHALLENGE

In the *Two Dimensional Random Walk*, a particle is bombarded by molecules. The particle moves one space up, down, right, or left, each with probabiliy of 1/4. Use **MTOSS** (see Exercise 12) to estimate how far the particle is likely to be from its initial position after 5000 bombardments.

Copyright © 1998 by Brendan Kelly Publishing Inc.

EXPLORATION 10 Monte Carlo Methods: Cards & Dice

BETTMANN

Stanislaw Ulam

HOW WOULD YOU COMPUTE THE PROBABILITY OF WINNING AT SOLITAIRE?

Stanislaw Ulam, one of the co-developers of the atomic bomb, was stationed in Los Alamos, New Mexico, during the 1940's. Between bomb tests, there was sufficient time for Ulam to engage himself in the game of solitaire. During one such involvement, he pondered the question,

In what fraction of all the games of solitaire will a player complete to the last card?.

That is, what is the probability of winning at solitaire? Since this problem is too complex (i.e. too many cases) to be solved using theoretical probability computations, Ulam programmed one of the early mainframe computers to simulate solitaire and he ran hundreds of trials. He then computed the proportion of times that the computer won. In this way he found the relative frequency of the "win" outcome and estimated its probability. Such a procedure for estimating the probability of an outcome by computing the experimental probability of that outcome in a simulation is called a *Monte Carlo technique.* The simulations in Exploration 9 that were used to calculate experimental probabilities are examples of Monte Carlo techniques. Such techniques have been in use prior to this century but it was Ulam, von Neumann, and others who formalized these methods and established them as a legitimate branch of mathematics.

The computer technology that has emerged in the last few decades has increased the power and scope of these methods to a hitherto unimagined extent. Some of the simple programs shown here for the TI-83 suggest new approaches to problems that were previously intractable.

WORKED EXAMPLE 1

Modify **DIETOSS** to create **MTOSS**, a program that creates the same kind of histogram as **DIETOSS** (see page 34), but for an M-sided die with all sides equally probable outcomes.

An employer discovers that 496 of the 2017 absences reported last month occurred on Fridays. Run **MTOSS** to estimate the probability that this would occur from illness alone, assuming that illness occurs on any of the five work days of the week with equal probability.

SOLUTION

Program **MTOSS** is shown on the right. We merely replace the number 6 (for 6 sides or faces) with the letter **M** (for **M** sides or faces) in program **DIETOSS**. We also change the definition of L_1 to give it dimension **M**.

Since sickness is equally likely to occur on any one of the 5 days, Monday through Friday, we can simulate the selection of a sick day by the toss of a 5-sided die with faces marked M, T, W, Th, and F. In our simulation, we toss the die 2017 times and observe whether outcome F occurs as often as 496 times. We repeat this experiment 20 times and record the fraction of times outcome F occurred 496 or more times. This gives us an estimate of the probability that this number of absences on Friday could occur by chance.

```
PROGRAM:MTOSS
:Input "HOW MANY TOSSES?",N
:Input "HOW MANY SIDES? ",M
:1→Xmin:M+1→Xmax
:0→Ymin
:Fix 0
:seq(Z,Z,1,M,1)→L₁
:0L₁→L₂
:For(I,1,N)
:randInt(1,M)→X
:L₂(X)+1→L₂(X)
:End
:max(L₂)+1→Ymax
:Plot1(Histogram,L₁,L₂)
:FnOff
:1→Xscl
:DispGraph
:
```


Copyright © 1998 by Brendan Kelly Publishing Inc.

Solution (cont'd)

When we run **MTOSS**, with **N** = 2017 (number of trials) and **M** = 5 (number of possible equally likely outcomes on each trial), we obtain a histogram like the one shown in the display.

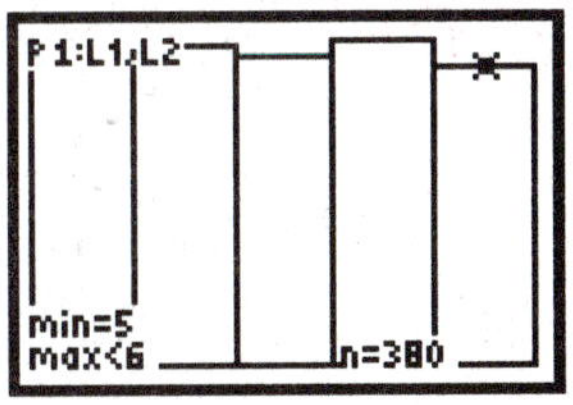

The frequency of outcome, "Friday" is represented by the rightmost bar on the histogram. When we press STAT ENTER we see from the table that on this replication of the simulation, only 380 sickdays fell on Friday. That is, on the first trial, the number of Friday sickdays was less than 496. To estimate the fraction of times the number of Friday sickdays will be less than 496, we run **MTOSS** 20 times. The number of Friday sickdays on replications 2 through 20 were: 389, 421, 395, 365, 370, 427, 385, 409, 406, 366, 389, 405, 430, 417, 401,

L1	L2	L3 1
1	421	------
2	410	
3	392	
4	414	
5	380	

L1(6)=

It seems that that there is a small probability that 496 sickdays would fall on Friday on the basis of chance alone. However, before introducing major policies to deal with these absences, management should collect data for several months to determine whether this pattern is maintained and to investigate other possible reasons for disproportionate absence on Fridays. We will investigate more formal tests of such hypotheses in later Explorations.

Worked Example 2

Each box of Brainfood Cereal contains a CD about one of the world's 10 most celebrated scientists. The 10 CDs that constitute a complete set are randomly distributed in equal quantities. What is the average (mean) number of boxes of cereal that a person could expect to buy in order to obtain the complete set of CDs?

Solution

How FULLSET simulates the purchase of cereal boxes.

Program **FULLSET** simulates the random selection of integers from the set of **N** integers, {1, 2, 3, … **N**}, where **N** is the number of CDs in a complete set. When integer, **X** is selected by the program, it is stored to $L_1(X)$, the Xth term of list L_1. We have a complete set when each integer between 1 and **N** has been selected at least once; that is, all terms of L_1 are non-zero. Until this occurs, the product of the terms of L_1 is zero, and the random drawing continues. The variable **S** counts the number of items selected. When the set is complete, i.e. **prod** $L_1 > 0$, the display shows, **S**, the number of selections required to get a full set.

```
PROGRAM:FULLSET
:0→S
:seq(0*X,X,1,N)→
L1
:While prod(L1)=
0
:randInt(1,N)→X
:X→L1(X)
:S+1→S
:End
:Disp "NUMBER OF
BOXES",S
:Return
:
```

How FULLSET is used

Before running **FULLSET**, we store 10 to the variable N, to indicate that there are 10 CDs in a complete set. When we ran **FULLSET** once, we obtained the display shown here. That is, on one replication of the simulation, it took 34 cereal boxes before we had the complete set of 10 CDs. Is this typical, or was this just an unlucky draw?

```
NUMBER OF BOXES
              34
            Done
```

To determine whether 34 is close to the average (mean) number of boxes we would have to purchase, we could run **FULLSET** 50 times and take the mean number of boxes required. Alternatively, we could write a program such as **FULLSETS** (shown here) to call **FULLSET** as a subroutine and replicate it **T** times.

```
PROGRAM:FULLSETS
:Disp "HOW MANY
IN A "
:Input "FULL SET
?",N
:Input "HOW MANY
TRIALS?",T
:0→R
:For(I,1,T)
:prgmFULLSET
:R+S→R
:End
:Disp "MEAN NUMB
ER "
:Disp "OF BOXES"
,R/T
:
```

When we run **FULLSETS**, we enter 10 and 20 in response to the prompts for the number of a full set and the number of trials. The display shows the final screen output for a single run of **FULLSETS**.

```
              38
NUMBER OF BOXES
              14
MEAN NUMBER
OF BOXES
              29
            Done
```

We conclude that on average, we would have to purchase about 30 boxes of cereal to get the full set of 10 CDs.

Copyright © 1998 by Brendan Kelly Publishing Inc.

Worked Example 3

a) What is the probability that two people in a group of N people have coincident birthdays; i.e. birthdays on the same day of the same month (but not necessarily the same year) ?

b) Enter and run program **COINCIDE** (with subroutine **BIRTH**) to estimate this probability when $N = 23$. Run **COINCIDE** for 200 trials to test your estimate.

Solution

a) Suppose we have a group of N people. We ask each person to write his or her birthdate on a slip of paper that we place in a hat. As we draw these slips of paper from the hat, we assume each of the 365 dates is equally likely (and we identify Feb. 29 with March 1).

We shall calculate first the probability that there are no coincident birthdays.
On the first draw there can be no coincidence. (It's the first one considered.)
On second draw, the probability that the date differs from the first: $\frac{364}{365}$

On third draw, the probability that the date differs from the first two: $\left(\frac{364}{365}\right)\left(\frac{363}{365}\right)$

•
•
•

On the Nth draw, the probability that the date differs from the previous (N-1) is:

$$\left(\frac{364}{365}\right)\left(\frac{363}{365}\right)\left(\frac{362}{365}\right)\cdots\cdots\left(\frac{365-(N-1)}{365}\right)$$

So the probability of coincidence is: $1-\left(\frac{364}{365}\right)\left(\frac{363}{365}\right)\left(\frac{362}{365}\right)\cdots\cdots\left(\frac{365-(N-1)}{365}\right)$

For the probability when N = 23, we substitute into the formula to obtain:

Probability of coincidence is $1-\left(\frac{364}{365}\right)\left(\frac{363}{365}\right)\left(\frac{362}{365}\right)\cdots\cdots\left(\frac{344}{365}\right)\left(\frac{343}{365}\right)$

To evaluate this expression on the TI-83, we use the **prod** and **seq** functions from the LIST menus and enter the expression shown in the display below.

```
1-prod(seq(X/365
,X,343,364)
            .507
```

To access the **prod(** command, press these keys:

2nd [LIST] ▶ ▶ 6

To access the **seq(** command, press these keys:

2nd [LIST] ▶ 5

The display shows that the probability of coincidence with 23 people ≈ 0.507—that is, coincidence is more likely than not!

```
COINCIDENCE AT
              16
FRACTION OF
TRIALS WITH
COINCIDENCE
             .52
            Done
```

To simulate the birthday problem, we entered both programs, **BIRTH** and **COINCIDE** and ran **COINCIDE** with N = 23 and **"NUMBER OF TRIALS"** = 200. We obtained the display on the left indicating a probability of coincidence of 0.520, which is close to the theoretical probability, 0.507.

```
PROGRAM:BIRTH
:seq(0*X,X,1,365
)→L1
:For(I,1,N)
:randInt(1,365)→
X
:If L1(X)≠0
:Then
:Disp "COINCIDEN
CE AT ",I
:S+1→S
:Return
:Stop
:End
:1→L1(X)
:End
:Disp "NO COINCI
DENCE"
:Return
```

```
PROGRAM:COINCIDE
:Input "HOW MANY
 PEOPLE? ",N
:Input "HOW MANY
 TRIALS? ",T
:Float
:0→S
:For(J,1,T)
:prgmBIRTH
:End
:Disp "FRACTION
OF"
:Disp "TRIALS WI
TH"
:Disp "COINCIDEN
CE",S/T
```


Copyright © 1998 by Brendan Kelly Publishing Inc.

Exercises

1. Describe what is meant by a Monte Carlo technique.

2. Each item on a 25-item multiple choice test offers 4 possible answers. Run **MTOSS** to estimate the probability that a student would score at least 10 correct if the student guesses randomly the answer to each item.

3. The probability that an archer will shoot into the inner circle of a target is 0.2. Estimate the probability that the archer will have at least 4 arrows out of 20 land within the inner circle.

4. At a crap table in a famous casino, the dice showed snake eyes (that is; both dice showed a single dot) 10 times on 120 rolls. Estimate the probability of this happening if all 36 outcomes on the roll of the two dice are equally probable?

5. In *Worked Example* 3, we discovered that the probability of ***no*** coincident birthdays in a group of 23 people is given by:

$$\left(\frac{364}{365}\right)\left(\frac{363}{365}\right)\left(\frac{362}{365}\right)\cdots\cdots\cdots\left(\frac{344}{365}\right)\left(\frac{343}{365}\right)$$

Write this expression using factorial notation.
Can you evaluate the expression using the PRB submenu of the MATH menu? Explain why or why not.

6. a) Run program **MTOSS** to simulate the probability that at least two people in a group of 5 people will have birthdays in the same month.

b) Calculate the theoretical probability that at least two people in a group of 5 people will have birthdays in the same month. Compare with the experimental probability in part a).

7. Modify program **BIRTH** to estimate the probability that at least two people in a group of N people will have birthdays in the same month. Call your modified program as a subroutine of **COINCIDE,** to estimate this probability using 50 trials.

8. Every weekday Professor Polyhedron leaves for the university at a random time between 7:00 A.M. and 7:15 A.M. The *Daily Planet* newspaper arrives at her home at a random time between 7:10 A.M. and 7:30 A.M. Estimate the probability that Professor Polyhedron receives her paper before leaving home on at least one day out of 5.

9. Enter programs **FULLSET** and **FULLSETS** into your calculator. Run **FULLSETS** to estimate the average (mean) number of packages of gum that a baseball fan might expect to purchase to collect the full set of 30 baseball cards. Assume that there is one baseball card in each package of gum and that all baseball cards are equally likely.

Investigations

10. a) Compute the probability of coincident birthdays for a group of N people where N takes each of these values:
(i) 10 (ii) 15 (iii) 20 (iv) 25 (v) 30 (vi) 35

b) Run **COINCIDE** to determine whether your answers in part a) are reasonable.

c) Create a table with L_1 taking the values 10 through 35 in increments of 5 and L_2 taking the corresponding probabilities of coincidence for these values of N.

d) Construct a line graph of L_2 vs. L_1.

11. a) Write a program, **TWODICE**, which simulates N tosses of a pair of dice. On each toss of the dice, **TWODICE** computes the total showing on the two dice and at the end of N tosses displays a histogram showing the frequency of each total between 2 and 12.

b) Run **TWODICE** for 100 tosses of the dice and record the experimental probability of each sum.

c) Compute the theoretical probability of each sum and compare with the experimental probability in part b).

12. a) We can run program **BINOMIAL** (see page 37) to simulate the tossing of 10 coins. On each toss, there are 11 possible outcomes: 0 heads, 1 head, 2 heads,...10 heads. Run **BINOMIAL** for 10,000 tosses of 10 coins.

b) Graph the function defined by $y = \frac{10000}{\sqrt{5\pi}} e^{-\frac{(x-5)^2}{5}}$

so that it overlays your histogram. What do you discover? Repeat this procedure for the tossing of 20 coins. Graph the function defined by

$$y = \frac{10000}{\sqrt{10\pi}} e^{-\frac{(x-10)^2}{10}}$$

Describe how the histogram changes as the number of coins increases.

CHALLENGE

The program **BINOMIAL** simulates M replications of the toss of N coins. Create a program, **DINOMIAL** which simulates T replications of the toss of N dice that are M-sided. (That is, each die has M faces, all of which are equally likely outcomes.) **DINOMIAL** should call MTOSS as a subroutine.

Run **DINOMIAL** to estimate the probability that at least two people in a group of 5 people will have birthdays in the same month. Compare this estimate with your answers to Exercises 6 and 7.

Copyright © 1998 by Brendan Kelly Publishing Inc.

Inferential Statistics – Prerequisite Concepts

A Random Variable

A *discrete random variable* defined on the set $\{x_1, x_2, x_3, \ldots x_n\}$ is a probability function that assigns to each x_i a probability $p(x_i)$, where $0 \le p(x_i) \le 1$, and $\sum_{i=1} p(x_i) = 1$. For example, if X is the random variable corresponding to the number of heads obtained on n tosses of a fair coin, then the probability that X takes the value k is given by $p(k) = {}_nC_k / 2^n$.

A *continuous random variable* X defined on the set of real numbers is a probability function that associates with each real value x a *probability density* $f(x)$, such that the total area under the graph of $y = f(x)$ is equal to 1. For example, if X is the random variable corresponding to the (Wechsler) I.Q. score of a randomly selected person, then the *probability density function* associated with x is $f(x) = \dfrac{1}{15\sqrt{2\pi}} e^{-\frac{1}{2}\left(\frac{x-100}{15}\right)^2}$. The function $F(x)$ such that $F(a) - F(b)$ is the area under the graph of $f(x)$ in the interval $a \le x \le b$, is called the *cumulative distribution function of* X. The cumulative distribution function $F(x)$ is important because $F(a) - F(b)$ is the probability that X takes a value in the interval $a \le x \le b$, for any real numbers a and b.

Percentiles

The pth *percentile of a discrete random variable* X is the number below which X will assume a value p% of the time.

The pth *percentile of a continuous random variable* X is the number such that X will assume a value below it p% of the time. More precisely, if $F(x)$ is the cumulative distribution function of X, then $F(p^{\text{th}} \text{ percentile}) = p/100$.

Confidence Intervals

If $X_{0.05}$ and $X_{0.95}$ denote the 5th and 95th percentiles respectively of a random variable X, then there is a 90% probability that a randomly selected value of X will lie in the interval $X_{0.05} \le X \le X_{0.95}$. This interval is a 90% *confidence interval* for the random variable X.

An appropriately constructed confidence interval contains, with a specified level of certainty, a particular population parameter such as a mean or standard deviation. For example, if $\bar{x}$ denotes the mean of a sample of size n taken from a normal population of mean μ and standard deviation σ, then $Z = \dfrac{\bar{x} - \mu}{\sigma / \sqrt{n}}$ is normally distributed with mean 0 and standard deviation 1. If $z_{0.025}$ denotes the 2.5th percentile of Z, then we assert that for a randomly chosen sample, there is a 95% probability that $-z_{0.025} \le \dfrac{\bar{x} - \mu}{\sigma / \sqrt{n}} \le z_{0.025}$. Solving for μ, we obtain

$$\bar{x} - z_{0.025}\,\sigma/\sqrt{n} \le \mu \le \bar{x} + z_{0.025}\,\sigma/\sqrt{n}$$

We can assert that 95% of all the intervals $(\bar{x} - z_{0.025}\,\sigma/\sqrt{n},\ \bar{x} + z_{0.025}\,\sigma/\sqrt{n})$ contain the true value of μ. This is a 95% *confidence interval for the estimation of* μ.

Copyright © 1998 by Brendan Kelly Publishing Inc.

UNIT 4 DISTRIBUTIONS & THEIR PROPERTIES

TO ACCESS THE DISTRIBUTIONS PRESS: **2nd** [DISTR]. We obtain the menu

```
DISTR DRAW
1:normalpdf(
2:normalcdf(
3:invNorm(
4:tpdf(
5:tcdf(
6:χ²pdf(
7:χ²cdf(
8:Fpdf(
9:Fcdf(
0:binompdf(
A:binomcdf(
B:poissonpdf(
C:poissoncdf(
D:geometpdf(
E:geometcdf(
```

If we press **2nd** [DISTR] **▶** we obtain

```
DISTR DRAW
1:ShadeNorm(
2:Shade_t(
3:Shadeχ²(
4:ShadeF(
```

SUMMARY OF COMMON DISTRIBUTIONS – EQUATIONS & COMMANDS

	Equation of the Probability Density Function	Probability Density Function Command	Cumulative Distribution Function Command
Binomial Distribution n trials, probability of success p	$\frac{n!}{k!(n-k)!}p^k(1-p)^{n-k}$	**binompdf(n, p, k)**	**binomcdf(n, p, k)**
Normal Distribution mean μ, standard deviation σ	$y=\frac{1}{\sigma\sqrt{2\pi}}e^{-\frac{1}{2}\left(\frac{x-\mu}{\sigma}\right)^2}$	**normalpdf(x, μ, σ)**	**normalcdf(L, U, μ, σ)** L-lower bound U-upper bound
***t* Distribution** n degrees of freedom	$t_n(x)=\frac{C_n}{\left(1+\frac{x^2}{n}\right)^{\frac{n+1}{2}}}$	**tpdf(x, n)** ↑ degrees of freedom	**tcdf(L, U, n)** L-lower bound U-upper bound n - degrees of freedom
χ^2 Distribution n degrees of freedom	$\chi_n^{\,2}(x)=\begin{cases}K_n x^{\frac{n-2}{2}}e^{-\frac{x}{2}} & \text{for } x>0\\ 0 & \text{elsewhere}\end{cases}$	**χ^2pdf(x, n)** ↑ degrees of freedom	**χ^2cdf(L, U, n)** L-lower bound U-upper bound n - degrees of freedom
***F* Distribution** n_1 and n_2 degrees of freedom	$f_{n_1,n_2}(x)=\begin{cases}\frac{k}{\left(1+\frac{n_1}{n_2}x\right)^{(n_1+n_2)/2}}x^{(n_1/2)-1} & \text{for } x>0\\ 0 & \text{elsewhere}\end{cases}$	**Fpdf(x, n_1, n_2)** ↑ ↑ degrees of freedom	**Fcdf(L, U, n_1, n_2)** ↑ ↑ degrees of freedom

EXPLORATION 11 The Binomial Distribution

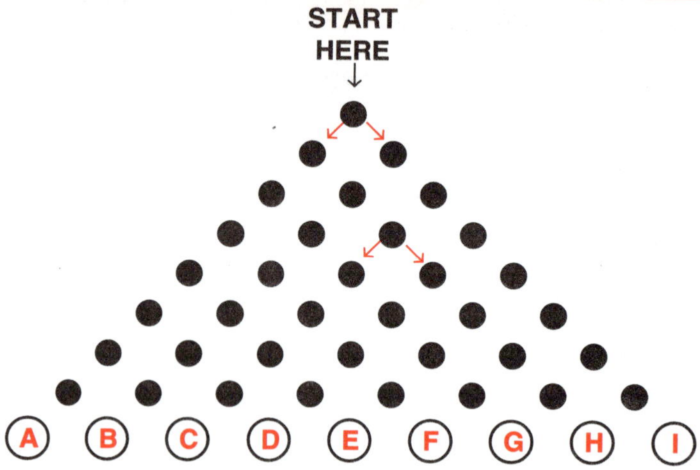

A computer game simulates an old pin-ball machine. A "ball" is launched at the post labelled "START HERE" and it moves with equal probability either southwest or southeast to an adjacent post in the second row. When it encounters that post it once more bounces in either direction with equal probability (southwest or southeast) to an adjacent post in the third row. This continues until it reaches the bottom row and terminates in one of the 9 slots, A through I. Conjecture an answer to each of the following questions.

- Which slot(s) will accumulate the most balls?
- Which slot(s) will accumulate the fewest balls?
- What is the probability that a ball will land in slot D?

Since the ball can move only downward to the left (southwest) or downward to the right (southeast), then we will represent these moves respectively by W and E. A path from the start to a slot can be represented as a string of 8 Ws and Es.

To end up in slot A, the ball must move only west; that is, its path must be WWWWWWWW. However, to end up in slot B, it must make 7 trips west and 1 trip east. Any permutation of 7Ws and 1E defines a path that ends in slot B. Similarly, any permutation of 6Ws and 2Es defines a path that ends in slot C.

The number of paths is the number of 8-letter strings of Ws and Es. Since each of the 8 letters can be chosen in 2 ways, the number of paths is 2^8; i.e., 256 paths. Since all paths are equally likely, each has a probability of 1/256.

All the paths that end in slot D can be represented as 8-letter strings consisting of 3Ws and 5Es. To count the number of such strings, we choose the 3 positions (out of the 8) into which Ws are to be placed in ${}_8C_3$ ways. The Es are assigned to the remaining positions. That is, the number of paths terminating in slot D is ${}_8C_3$. The probability for *each* path is 1/256, so the probability that a ball will land in slot D is ${}_8C_3/256$ or 7/32. Can you deduce the probability that a ball will terminate in slot E? Using the symmetry of the array, what can you conclude about the probabilities for the other slots?

The set of probabilities corresponding to each outcome is called the *binomial distribution* because the numbers ${}_8C_r$ are precisely the coefficients in the binomial expansion of $(1 + x)^8$.

```
■ expand((1 + x)^8)
  x^8 + 8·x^7 + 28·x^6 + 56·x^5 + 70·x^4 + 56·x^3 + 28·x^2 + 8·x + 1
expand((1+x)^8)
MAIN          RAD AUTO          3RAD AUTO          3D   1/30
```

←This is a TI-92 screen display showing the expansion of $(1 + x)^8$

Copyright © 1998 by Brendan Kelly Publishing Inc.

If we write the number of ways of reaching each post in the computer game on page 46, we obtain the following array.

$$
\begin{array}{ccccccccccccccc}
 & & & & & & & 1 & & & & & & & \\
 & & & & & & 1 & & 1 & & & & & & \\
 & & & & & 1 & & 2 & & 1 & & & & & \\
 & & & & 1 & & 3 & & 3 & & 1 & & & & \\
 & & & 1 & & 4 & & 6 & & 4 & & 1 & & & \\
 & & 1 & & 5 & & 10 & & 10 & & 5 & & 1 & & \\
 & 1 & & 6 & & 15 & & 20 & & 15 & & 6 & & 1 & \\
1 & & 7 & & 21 & & 35 & & 35 & & 21 & & 7 & & 1
\end{array}
$$

Observe that this triangle of numbers has some special properties:

- Each number in the triangle is the sum of the two adjacent numbers in the row above it. (Can you explain this in terms of the number of paths?)
- The numbers are symmetrical about a vertical line through the center.
- The sum of the numbers in the nth row is 2^n.

What patterns can you see in the diagonal lines?

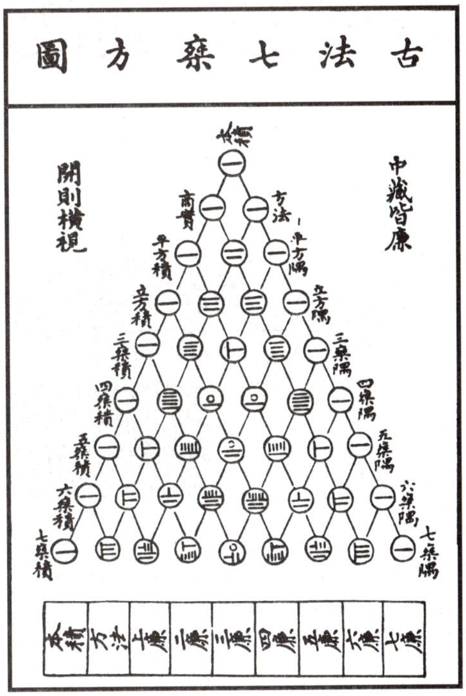

This reproduction appeared in 1303 A.D. in Chu SHIH-Chieh's Ssu Yuan Yii Chien. In 1665, French mathematician Blaise Pascal published a treatise in which he presented this array in a slightly different form. It has subsequently been called *Pascal's Triangle.*

In our investigation of the computer simulated pin-ball machine on page 46, we saw that the number occupying the $(k+1)^{th}$ position in the nth row is the number of ways of choosing k out of n positions into which Ws are to be placed; i.e. ${}_nC_k$.

Since the sum of the numbers in the nth row is 2^n, then there are a total of 2^n paths that terminate in that row. Since all paths are equally likely, the probability that a ball will reach the $(k+1)^{th}$ position in the nth row is ${}_nC_k / 2^n$. That is,

When the selection of Ws and Es are equally likely, the probability of selecting exactly k Ws on n selections is:

$$\frac{n!}{k!(n-k)!}\left(\frac{1}{2^n}\right)$$

This formula gives the probability of exactly k occurrences of a particular outcome on n trials when the probability of that outcome is 1/2. But what is the probability when this outcome has some probability p where $p \neq 1/2$?

We can generalize the argument given above to obtain the following result.

If the probability that an event will occur on any trial is p, then the probability that it will occur *exactly* k times on n trials is given by

$$\frac{n!}{k!(n-k)!}p^k(1-p)^{n-k}$$

The general formula below applies to repeated trials of an experiment in which:

- there are exactly two possible outcomes.
- each trial is independent of the others.
- the probability of each outcome is the same for each trial.

Replications of experiments satisfying these three conditions are called *Bernoulli Trials.*

Copyright © 1998 by Brendan Kelly Publishing Inc.

Worked Example 1

If both sexes are equally likely at birth, what is the probability that exactly 3 of 5 children will be girls?

Solution

In *Worked Example* 2, p. 37, we showed how the simulation **COINTOSS** could be used to estimate this probability.

The formula presented on page 47 enables us to move directly to the solution by observing that we can regard the birth of a child as a trial (as the mother will attest during delivery) in which the outcome "girl" has probability 1/2. The birth of 5 children can be taken as 5 trials and substitution of $n = 5$ and $k = 3$ into the formula yields the probability of 3 "girl" outcomes on these 5 trials as:

$$\frac{5!}{3!\,2!}\left(\frac{1}{2}\right)^5 \text{ or } \frac{5}{16}.$$

This theoretical probability of 0.3125 is close to the experimental probability of 0.29 that we obtained in *Worked Example* 3 on page 37 using 100 trials.

Note: Your TI-83 has a binomial probability density function, **binompdf(**, which can be used to calculate this probability directly.

probability of a girl ↓

```
binompdf(5,.5,3)
                .3125
```

← number of girls; ↑ number of trials

To access **binompdf(** press:

2nd [DISTR] **0**

Worked Example 2

About 8% of all rental cars from a particular rental agency are defective. What is the probability that a random sample of 15 cars will have:

a) exactly 2 defective cars?

b) at most 2 defective cars?

Solution

a) The probability of selecting a defective car is about 0.08. The selection of 15 cars can be regarded as 15 trials in which the probability of outcome "defective" is 0.08. The probability that there will be exactly 2 "defectives" in 15 trials is obtained by substituting $n = 15$, $k = 2$, and $p = 0.08$ into the formula at the bottom of page 47. We obtain

$$\frac{15!}{2!\,13!}(0.08)^2(0.92)^{13} \text{ or } 0.227$$

That is, the probability that there will be exactly 2 defective cars is about 0.227. ←

```
binompdf(15,.08,
2)
         .2273060675
```

b) Proceeding as in part a), we find that the probability that there will be exactly one defective car is $\frac{15!}{1!\,14!}(0.08)^1(0.92)^{14}$ or 0.373

The probability that there will be no defective cars is $(0.92)^{15}$ or 0.286

The probability that there will be at most 2 defective cars is: p(0) + p(1) + p(2), where p(0), p(1), and p(2) denote respectively the probabilities that there will be exactly 0, 1, or 2 defective cars. Therefore, the probability that there will be at most 2 defective cars is: 0.286 + 0.373 + 0.227 = 0.886.

We observe that it is almost certain (0.937) that a selection of 15 cars will have at most 3 that are defective.

Note: In this application, we have assumed that the defective cars are randomly distributed throughout a large population of rental cars. This enables us to assume that the "trials" are independent.

Note: Your TI-83 has a binomial cumulative density function, **binomcdf(**, which can be used to calculate this probability directly.

number of trials ↓

```
binomcdf(15,.08,
2)
         .8870348686
```

← probability of a defective car; ↑ cumulative upper limit

To access **binompdf(** press:

2nd [DISTR] **ALPHA** [A]

Copyright © 1998 by Brendan Kelly Publishing Inc.

Exercises

1. Explain the meaning of each term.
   a) binomial distribution
   b) independent trials
   c) Bernoulli trials

2. What is the probability that a student will guess exactly 10 questions correctly on a 20-item true-false test?

3. Consider the following problem:

What is the probability of obtaining exactly 5 spades on the random draw of 8 cards from a standard deck of 52 cards?

A student answers as follows:
We must calculate the probability of drawing 5 spades in 8 trials. Since there are 13 spades in a deck of 52 cards, the probability of drawing a spade on any draw is 1/4. Therefore the probability of drawing exactly 5 spades on 8 draws is obtained by substituting $n = 8$, $k = 5$ and $p = 1/4$ into the formula at the bottom of page 47. We obtain

$$\frac{8!}{5!\,3!}(0.25)^5(0.75)^3 \text{ or } 0.023$$

Is the student's solution correct? Explain why or why not.

4. a) Use the binomial formula to calculate the probability that there will be exactly 5 boys in a family of 7 children.
b) Describe at least two assumptions which you are making when you apply this formula.

5. Each item on a 25-item multiple choice test offers 4 possible answers. Calculate the probability that a student would score exactly 13 correct if the student guesses randomly the answer to each item.

6. The probability that an archer will shoot into the inner circle of a target is 0.2. Calculate the probability that the archer will have at least 4 arrows out of 20 land within the inner circle.

7. Assume that all months are equally likely birth months.

a) Calculate the probability that exactly two people in a group of 5 people will have birthdays in January.

b) Calculate the probability that at least two people in a group of 5 people will have birthdays in January.

c) Calculate the probability that at least two people in a group of 5 people will have birthdays in the same month.

d) Is your answer to part c) twelve times the probability you calculated in part b)? Explain why or why not.

8. Every weekday Professor Polyhedron leaves for the university at a random time between 7:00 A.M. and 7:15 A.M. The *Daily Planet* newspaper arrives at her home at a random time between 7:10 A.M. and 7:30 A.M. Estimate the probability that Professor Polyhedron receives her paper before leaving home on at least one day out of 5.

Investigations

AND WHEN WE TOSS AN INFINITE NUMBER OF COINS???

9. The binomial formula given on page 47 indicates that the probability of obtaining *exactly* k heads on the toss of n coins is given by:

$$\frac{n!}{k!(n-k)!}\left(\frac{1}{2^n}\right)$$

We denote by $p(k)$ the probability of achieving exactly k heads on the toss of n coins as a function of k when n is fixed.

$$p(k) = \frac{n!}{k!(n-k)!}\left(\frac{1}{2^n}\right)$$

$p(k)$ gives a probability corresponding to each positive integral value of k where $k \leq n$. We say that $p(k)$ defines the *binomial distribution.* How does the probability of obtaining exactly k heads change as the number of coins tossed becomes very large?; that is, how does the binomial distribution change as $n \to \infty$? To find out, complete this investigation.

a) To graph the probability density function for the binomial distribution with 10 coins, define L_1 and L_2 as shown in the displays.

```
seq(X,X,0,10)→L1
```

```
binompdf(10,.5,L
1)→L2
```

Then create a data plot like the one in the display showing the *xy*-Line with L_1 and L_2 as the Xlist and Ylist respectively. Use these window settings:
$0 \leq x \leq 40;\ 0 \leq y \leq 0.3.$

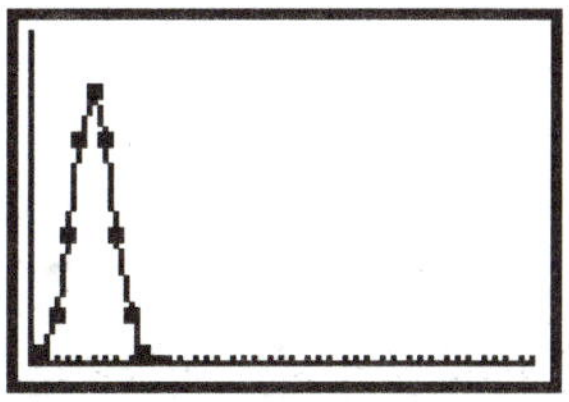

Trace along the *xy*-Line to find the probability that there will be exactly 5 heads on the toss of 10 coins.

Proceed as above to graph the binomial distributions corresponding to the tossing of 30 and 50 coins,

b) Describe how the binomial distribution changes as n increases.

c) When n coins are tossed, what number of heads is most likely to occur?

d) Stirling's approximation to $n!$ as n becomes large is:

$$n! \sim \sqrt{2\pi n}\left(\frac{n}{e}\right)^n \quad \text{The symbol} \sim \text{means "approaches close to".}$$

Use this formula to approximate the probability that exactly $n/2$ heads will occur on the toss of n coins as n becomes large.

Note: The displays above can also be generated by graphing in sequence mode.

Copyright © 1998 by Brendan Kelly Publishing Inc.

EXPLORATION 12 The Normal Distribution

In *Exploration* 11 we learned that the probability of obtaining *exactly* k heads on the toss of n coins is given by:

$$p(k) = \frac{n!}{k!(n-k)!}\left(\frac{1}{2^n}\right)$$

This set of probabilities corresponding to each outcome is called the *binomial distribution* because the numbers ${}_nC_k$ are precisely the coefficients that appear in the terms of the binomial expansion of $(1+x)^n$.

Frequency of Heads	Theoretical Probability	
0	${}_{10}C_0 \bullet 2^{-10}$	$\approx 9.76 \times 10^{-4}$
1	${}_{10}C_1 \bullet 2^{-10}$	$\approx 9.76 \times 10^{-3}$
2	${}_{10}C_2 \bullet 2^{-10}$	≈ 0.044
3	${}_{10}C_3 \bullet 2^{-10}$	≈ 0.117
4	${}_{10}C_4 \bullet 2^{-10}$	≈ 0.205
5	${}_{10}C_5 \bullet 2^{-10}$	≈ 0.246
6	${}_{10}C_6 \bullet 2^{-10}$	≈ 0.205
7	${}_{10}C_7 \bullet 2^{-10}$	≈ 0.117
8	${}_{10}C_8 \bullet 2^{-10}$	≈ 0.044
9	${}_{10}C_9 \bullet 2^{-10}$	$\approx 9.76 \times 10^{-3}$
10	${}_{10}C_{10} \bullet 2^{-10}$	$\approx 9.76 \times 10^{-4}$

The program **BINOMIAL**, presented on page 37, simulates m replications of the toss of n coins. When we ran **BINOMIAL** for M = 10 000 and N = 10 (that is, 10 000 tosses of 10 coins) we obtained the histogram shown below for the frequency of each of the 11 possible outcomes, i.e.: 0 heads, 1 head, 2 heads…10 heads. The experimental probabilities corresponding to these outcomes are shown in column L_3 of the table below. They are obtained by merely dividing these relative frequencies by 10,000. Observe how closely these approximate the theoretical probabilities in the table opposite.

↑ **L_2: Frequency of Each Outcome on 10,000 tosses $0 \le y \le 2412$**

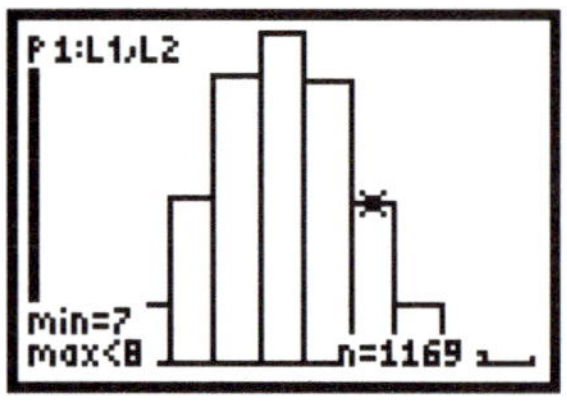

L_1: Number of Heads on the toss of 10 coins $0 \le x \le 11$

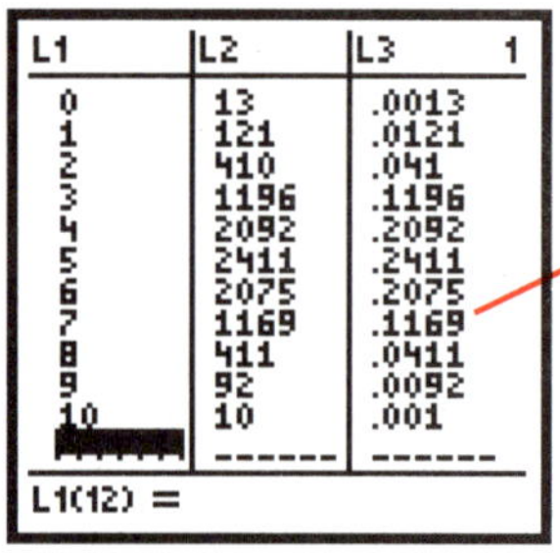

L1	L2	L3
0	13	.0013
1	121	.0121
2	410	.041
3	1196	.1196
4	2092	.2092
5	2411	.2411
6	2075	.2075
7	1169	.1169
8	411	.0411
9	92	.0092
10	10	.001

L1(12) =

We observe in the histogram that the most likely number of heads is 5 and the probabilities decrease as the balance between the frequencies of heads and of tails decreases. The histogram is virtually symmetric about the outcome "5". The *xy*-Line showing the experimental probabilities L_3 vs. the outcomes L_1 is shown in the display below left. The area under this curve is 1 corresponding to the total probability of all the outcomes.

xy-Line for 10,000 Tosses of 10 Coins

↑ Probability of each integral value of x

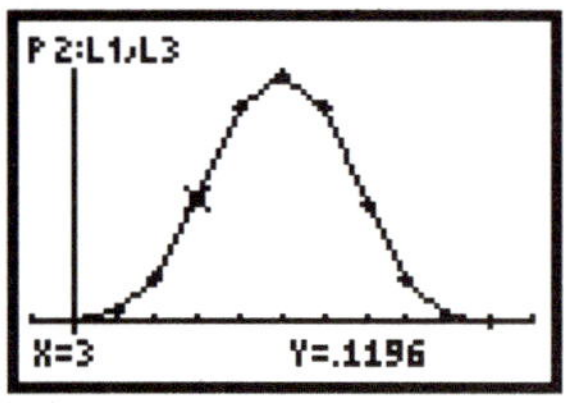

Graph of the normal distribution

↑ Probability density of each value of x

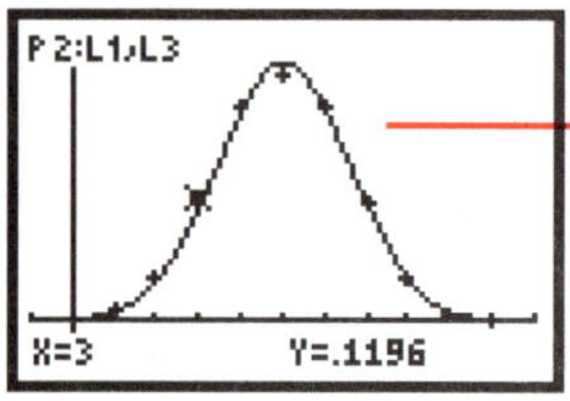

$$y = \frac{1}{\sqrt{5\pi}} e^{-\left(\frac{x-5}{5}\right)^2}$$

As the number of coins tossed is increased, the corresponding *xy*-Lines approach the shape of the so-called *normal curve* shown in the display above right. When this curve describes the probability densities of values of a continuous variable, it is the graph of a *normal distribution.*

A complete mathematical representation of the normal distribution was achieved when Abraham DeMoivre in the 1730s discovered that the equation of the curve corresponding to a normal distribution with mean μ and standard deviation σ is:

$$y = \frac{1}{\sigma\sqrt{2\pi}} e^{-\frac{1}{2}\left(\frac{x-\mu}{\sigma}\right)^2}$$

The constant $\frac{1}{\sigma\sqrt{2\pi}}$ was chosen so that the area under the curve is one square unit.

Copyright © 1998 by Brendan Kelly Publishing Inc.

In the late 18th century, Karl F. Gauss (see *Exploration* 7) discovered that repeated measurements of the same object gave rise to a histogram with the general shape of the normal curve. Since Gauss was one of the important pioneers in the study of the normal distribution, it is sometimes called the *Gaussian* distribution.

In the mid 19th century, Sir Francis Galton demonstrated, through extensive data collection, that many human characteristics such as height and I.Q. are distributed throughout the entire population in accordance with the normal distribution. For this reason and for its special mathematical properties, the normal distribution has become the basis of most studies involving human populations.

Worked Example 1

Galton recorded the heights of 8585 men in Great Britain. The *envelope* or outline of the histogram corresponding to these heights is a normal curve with mean $\mu = 67$ inches and standard deviation $\sigma = 2.5$ inches.

a) Graph the normal curve corresponding to this data and estimate the proportion of people who were between 66 and 68 inches tall.

b) What fraction of the adult male population in Great Britain was within one standard deviation of the mean height?

Solution

a) We substitute $\mu = 67$ and $\sigma = 2.5$ into the general equation for the normal distribution (given on page 50) to obtain:

$$y = \frac{\sqrt{2}}{5\sqrt{\pi}} e^{-2\left(\frac{x-67}{5}\right)^2}$$

$50 \le x \le 80;$
$0 \le y \le 0.2$

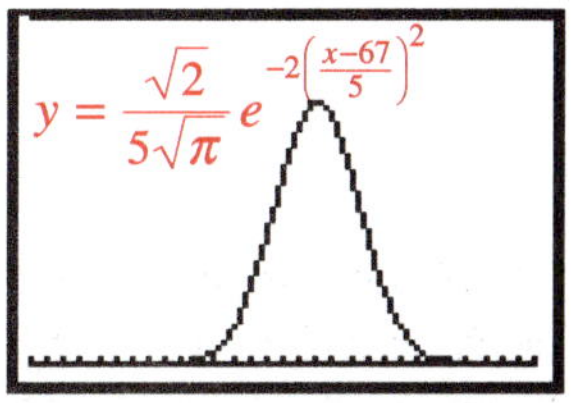

To graph this equation, we observe that the maximum height occurs at $x = 67$; i.e., at the mean value of x. For this value of x, $y = \frac{\sqrt{2}}{5\sqrt{\pi}} \approx 0.16$. Furthermore, the heights of most men will lie between 50 and 80 inches, so we set the window to: $50 \le x \le 80$ and $0 \le y \le 0.2$. We press **Y =** , enter the equation above, and press **GRAPH** to obtain the graph in the display.

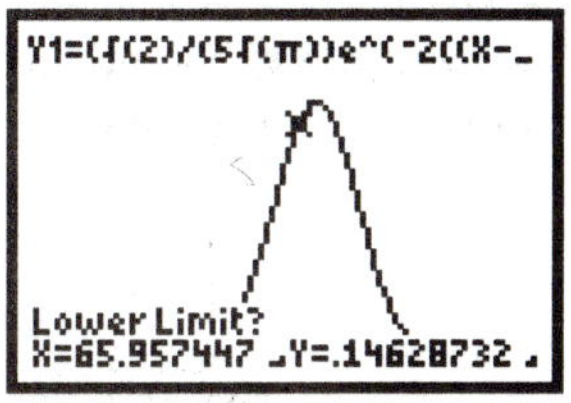

To determine the proportion of men whose heights were between 66 and 68 inches, we must find the area under the normal curve that lies in the interval: $66 \le x \le 68$. (This is analogous to adding the areas under the xy-Line to total the probabilities that the number of outcomes lies in a particular interval.) To find this area we will use the definite integral function, which is selection 7 on the **2nd** [CALC] menu. If you have not studied calculus, all you need to know is that the symbol $\int f(x)dx$ means "the area under the curve".
When we press: **2nd** [CALC] **7** , we obtain a graph and the prompt, **Lower Limit?** as shown in the display. In response to the prompt, we enter 66 followed by **ENTER** . We obtain the prompt **Upper Limit?** as shown in this display.

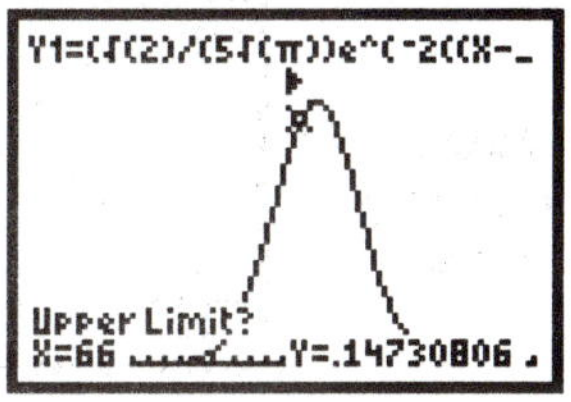

In response to the prompt, **Upper Limit**, we enter 68 followed by **ENTER** . We obtain the bottom display. This display indicates that the area under the normal curve and in the interval $66 \le x \le 68$ is about 0.31. That is, about 31.0% of the men were between 66 and 68 inches tall.

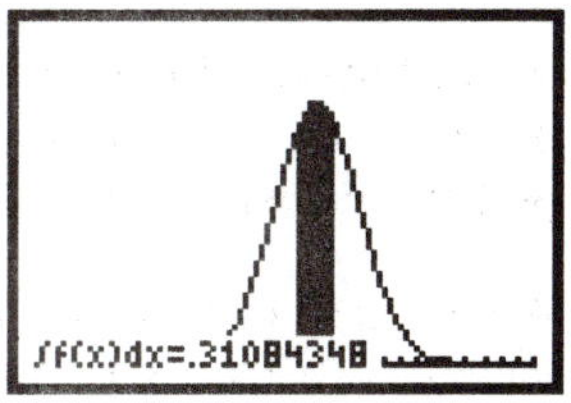

b) To determine the proportion of men whose heights lie within one standard deviation of the mean, we proceed as in part a) to determine the area under the curve between $67 - 2.5$ and $67 + 2.5$. Using the window $60 \le x \le 70$; $0 \le y \le 0.2$ we find that about 68% of the men were within one standard deviation of the mean.

Copyright © 1998 by Brendan Kelly Publishing Inc.

In *Worked Example* 1, we entered the equation for the normal distribution with mean 67 and standard deviation 2.5 to analyze Galton's data. Since the normal distribution is used so frequently in statistical analysis, the command **ShadeNorm(** has been embedded in your TI-83. *Worked Example* 2 shows how this command can be used.

WORKED EXAMPLE 2

a) Solve *Worked Example* 1 using the TI-83 **ShadeNorm(** command.

b) Determine the proportion of men who were more than 70 inches tall.

SOLUTION

a) To graph the normal probability density function with $\mu = 67$ and $\sigma = 2.5$ and shade the area under the curve in the interval $66 \le x \le 68$, we press:

2nd [DISTR] ▶ ENTER to obtain the command **ShadeNorm(**.

Then we enter the lower bound, upper bound, mean, and standard deviation as shown in the display. Before executing this command, we set the window to:

$$57 \le x \le 77;\ 0 \le y \le 0.2$$

Then we press ENTER to obtain the graph shown in the display. The display shows that the area under the curve and between 66 and 68 is 0.310... Since the total area under the curve is 1, we conclude as in *Worked Example* 1 that about 31.0% of the men were between 66 and 68 inches tall.

```
ShadeNorm(66,68,
67,2.5)
```

lower bound ↓; upper bound ↑; ↑ μ; ↑ σ

Area between 66 and 68 and under the normal curve with $\mu = 67$, $\sigma = 2.5$

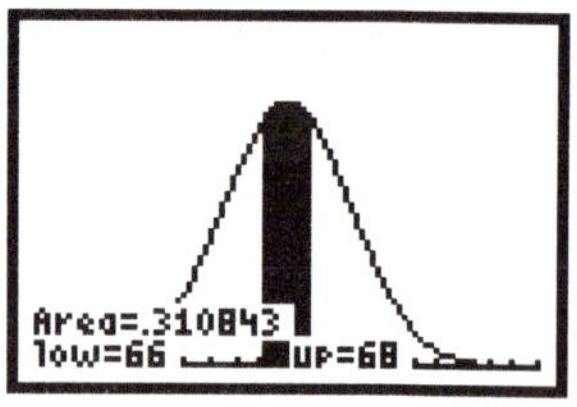

The proportion of the population that lies within the one standard deviation of the mean is the proportion in the interval $67 - 2.5 \le x \le 67 + 2.5$; i.e. $64.5 \le x \le 69.5$. The command **ShadeNorm(64.5,69.5,67,2.5)** yields an area 0.682...confirming *Worked Example* 1.

b) The proportion of men in the interval $x \ge 70$ is obtained by entering the command shown in the display. (Note that we use **E99** for ∞ and **-1E99** for –∞.) This yields an area of 0.11507, indicating that about 11.5% of the men were more than 70 inches tall.

```
ShadeNorm(70,E99
,67,2.5)
```

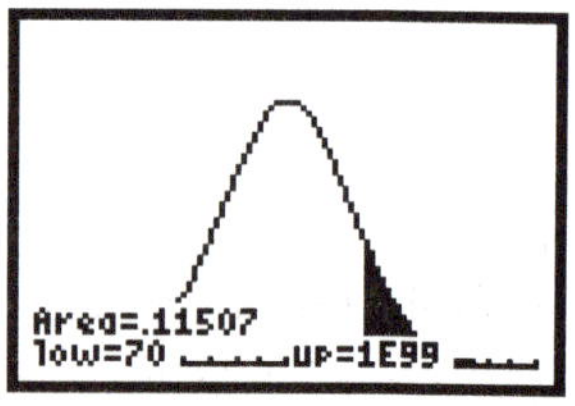

The probability density function and the cumulative distribution function of the normal distribution are embedded in the TI-83 commands **normalpdf(** and **normalcdf(** respectively. To find the probability density of a normal distribution with mean μ and standard deviation σ at any point x_0, we enter **normalpdf(**x_0**,** μ**,** σ**)**, filling in the numerical values for x_0, μ and σ. To determine the area under this normal curve and in the interval $x_{min} \le x \le x_{max}$, we enter **normalcdf(**x_{min}**,** x_{max}**,** μ**,** σ**)**. This command is like **ShadeNorm(**, except it does not graph the function. To graph the probability density function of the normal distribution with mean μ and standard deviation σ, we paste **normalpdf(**x**,** μ**,** σ**)** to the **Y=** editor , filling in the values for μ and σ.

To access the menu with commands **normalpdf(**, **normalcdf(** and **invNorm(** press:

2nd [DISTR]

WORKED EXAMPLE 3

The xth percentile in the Galton study is the height below which x% of the population registers. Determine the height corresponding to the 95th percentile in the Galton study. Use the **normalcdf(** command to verify your answer.

SOLUTION

The command **invNorm(** x, μ, σ**)** yields the xth percentile of the normal distribution with mean μ and standard deviation σ, corresponding to an input value of x. The display shows that 95% of the population in the Galton study was shorter than 71.112 inches. To verify that 95% of the area under this normal curve is to the left of $x = 71.112$, we evaluate **normalcdf(-1E99,71.112,67,2.5)** to obtain 0.95 as shown in the display.

```
invNorm(.95,67,2
.5)
     71.11213406
```

```
normalcdf(-1E99,
71.112,67,2.5)
     .9499944845
```


Copyright © 1998 by Brendan Kelly Publishing Inc.

Exercises

1. Write the equation of the normal curve corresponding to the normal distribution with:
a) $\mu = 50, \sigma = 10$ b) $\mu = 60, \sigma = 10$ c) $\mu = 80, \sigma = 10$

2. Graph each of the normal curves in Exercise 1 in the range: $0 \le x \le 100$; $0 \le y \le 0.05$. How does the graph of a normal distribution change if μ is increased by k units and σ is unchanged?

3. Write the equation of the normal curve corresponding to the normal distribution with:
a) $\mu = 50, \sigma = 5$ b) $\mu = 50, \sigma = 10$ c) $\mu = 50, \sigma = 20$

4. Graph each of the normal curves in Exercise 3 in the range: $0 \le x \le 100$; $0 \le y \le 0.1$. How does the graph of a normal distribution change if σ is increased by a factor k and μ is unchanged? Describe the relationship between the dispersion of a set of normally distributed data and the shape of the corresponding normal curve.

5. a) The wavelength of the light emitted by a particular isotope of phosphorus is measured 100 times. Why might you expect the measurements to be symmetrically distributed about the mean?

b) Prove algebraically that a normal curve is symmetric about the line $x = \mu$.

6. The I.Q. scores of Americans on the Wechsler test of intelligence are normally distributed with a mean $\mu = 100$ and a standard deviation $\sigma = 15$.

a) Write the equation of the corresponding normal curve that gives the probability density corresponding to each score.
b) Graph your equation using appropriate window settings.
c) Determine the percentage of people who have I.Q. scores between: (i) 85 and 115 (ii) 90 and 100.
d) When the Wechsler test was administered to a large random sample of people, N of them scored exactly 100. What score would you expect to be attained by exactly $\frac{N}{2}$ people?

7. In a recent study of the heights of 100,000 adult females throughout the world, it was found that the height, h, is a normally distributed variable with mean, $\mu = 162.5$ cm and standard deviation, $\sigma = 6.1$ cm.

a) Write the equation of the corresponding normal curve that gives the probability density for each height.
b) Graph your equation using appropriate window settings.
c) Estimate the percentage of adult females who are less than 160 cm in height.
d) Using the definition of percentile in *Worked Example* 3, determine the height of a woman who is at the 95th percentile.

Investigations

8. In *Exploration* 1, we presented the histogram of SAT math scores for the males in 1990. We can approximate these data by a normal distribution with mean 504 and standard deviation 125.

a) Graph the probability density function for this normal distribution in the window $200 \le x \le 800$; $0 \le y \le 0.004$.

b) About 75.6% of all males scored below 600 on the SAT math test. Use the normal distribution with mean 504 and standard deviation 125 to check this value. How close an approximation to the true value is provided by the normal distribution?

c) Use the normal distribution to estimate the percent of males who scored 700 or above. How does this estimate compare with the actual percentage of 6.53%?

d) Use the normal distribution to estimate the percentage of males who scored in the range $400 \le x < 600$. Compare this estimate with the actual percentage of 52.7%.

e) Estimate the score which corresponds to the 95th percentile. Does this estimate support your answer in part c) above?

f) Find the value of δ such that 68% of all the males scored in the interval $504 - \delta \le x \le 504 + \delta$.

9. The graph below shows the area under a normal curve with $\mu = 0$ for various intervals of width one standard deviation.

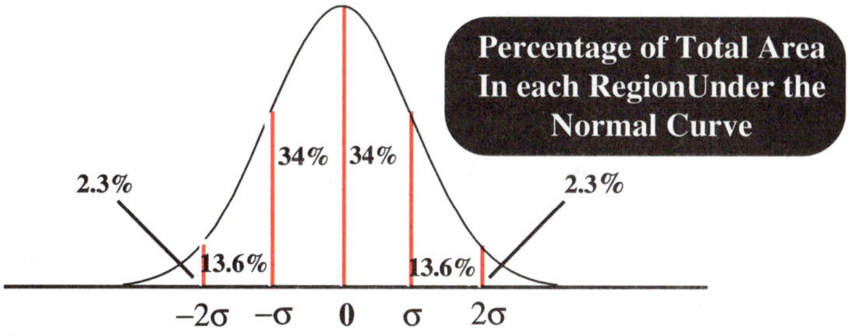

This graph shows that for *all* normal distributions, no matter what mean or standard deviation, 68% of the area under the curve lies within one standard deviation of the mean. Verify this is true for all the normal curves in Exercise 3.

ARE YOU GIFTED?

A school board identifies a student as *gifted* if she scores more than two standard deviations above the mean on the Wechsler test of intelligence. (See Exercise 6.)

a) What score on the Wechsler test is used as the criterion for identifying a gifted student?

b) About how many students could the board expect to have in its gifted program if there are 48,000 students in total?

Copyright © 1998 by Brendan Kelly Publishing Inc.

EXPLORATION 13 The t Distribution

While working as an executive for the Guinness Brewery, W.S. Gossett (1876-1936) developed a statistical method for analyzing quality control data in cases where the samples are small. Since Guinness did not want the other breweries to realize the value of this technique, Gossett published his discovery under the pseudonym "Student". Gossett's contribution was the discovery of a special continuous distribution that is used to estimate the mean of a population when its standard deviation is unknown and only a small sample of values is known. As a tribute to Gossett, this is sometimes called the *Student's distribution,* although today it is most often called the *t* distribution. The test that uses the *t* distribution is referred to as a *t*-test. It is somewhat ironic that a test developed in a brewery would come to be called a *t*-test. This is further evidence of the wry wit for which mathematicians are known.

The *t*-distribution of n degrees of freedom has probability density function $t_n(x)$ given by:

$$t_n(x) = \frac{C_n}{\left(1+\frac{x^2}{n}\right)^{\frac{n+1}{2}}}$$

← where C_n is a constant such that the area under the graph of the probability density function is 1.

$-\infty < x < \infty$

Worked Example 1

a) Find the values of the constant C_n for the t distributions of $n = 1$, 3, and 30 degrees of freedom.

b) Graph $t_1(x)$, $t_3(x)$, and $t_{30}(x)$ in the window $-3 \le x \le 3$; $0 \le y \le 0.5$.

c) Determine the area under the graph of $t_3(x)$ and in the interval $-1.6 \le x \le 1.6$, and interpret what this means.

Solution

a) Substitution of $n = 1$ into the expression for $t_n(x)$ yields $t_1(x) = \frac{C_1}{1+x^2}$.

To find the value of C_1, we graph $Y_1(x) = \frac{1}{1+x^2}$ in the window $-30 \le x \le 30$; $0 \le y \le 1.2$.

Then press **2nd** [CALC] **7** and enter –30 and 30 in response to the prompts, **Lower Limit?** and **Upper Limit?** followed by **ENTER**. We obtain the display shown here, indicating that the area under the curve is about 3.075. To make the area under the graph of $t_1(x)$ equal to 1, we must divide by 3.075; i.e. set $C_1 = 3.075^{-1} \approx 0.32$.
Proceeding as above, we find $C_3 \approx 0.367$ and $C_{30} \approx 0.395$.

The area under the graph of $(1 + x^2)^{-1}$ in $-30 \le x \le 30$

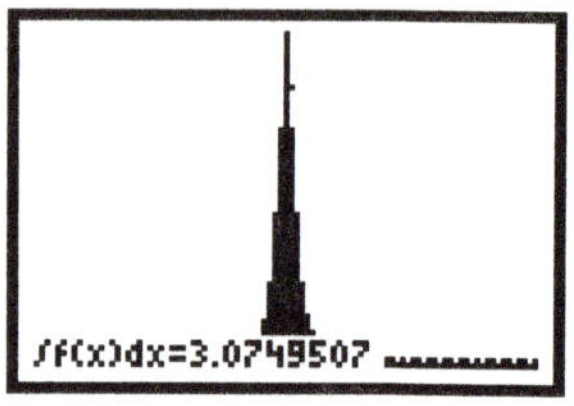

b) Substituting the values of C_n obtained in part a), we have:

$$t_1(x) \approx \frac{0.32}{1+x^2},\quad t_3(x) \approx \frac{0.367}{\left(1+\frac{x^2}{3}\right)^2} \quad \text{and} \quad t_{30}(x) \approx \frac{0.395}{\left(1+\frac{x^2}{30}\right)^{\frac{31}{2}}}$$

Entering these functions into the Y= editor and graphing yields the display on the right.

The *t*-distributions with 1, 3 and 30 degrees of freedom

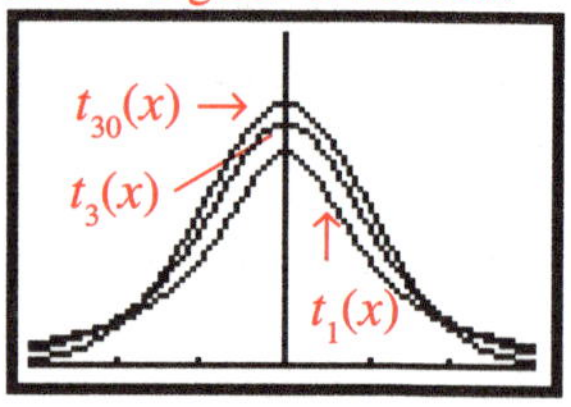

c) To determine the area under the graph of $t_3(x)$ in the interval $-1.6 \le x \le 1.6$, we press:

2nd [CALC] **7** and enter –1.6 and 1.6 in response to the prompts, **Lower Limit?** and **Upper Limit?**. Then we press **ENTER** and obtain the display shown here. We see that the contained area is 0.7909. That is, the probability that a random variable having a t distribution with 3 degrees of freedom takes a value between –1.6 and 1.6 is about 0.79.

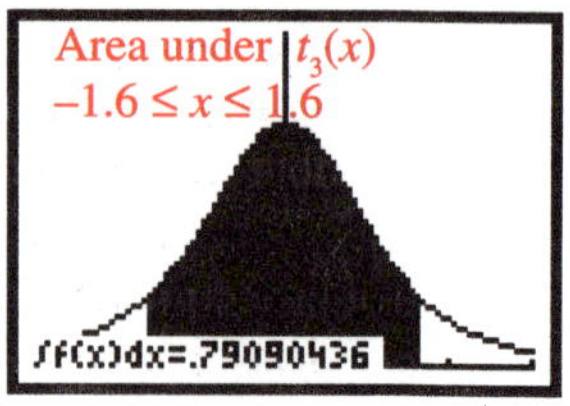

Copyright © 1998 by Brendan Kelly Publishing Inc.

To further reduce computational complexities, your TI-83 has the definition of the t distribution embedded in its memory. In the example that follows, we show how to use the TI-83 commands, **tpdf(** (the t distribution probability density function) and **Shade_t(** to solve *Worked Example* 1.

WORKED EXAMPLE 2

a) Graph $t_1(x)$, $t_3(x)$ and $t_{30}(x)$ in the window $-3 \le x \le 3$; $0 \le y \le 0.5$. Evaluate C_1, C_3, and C_{30}.

b) Determine the area under the graph of $t_3(x)$ and in the interval $-1.6 \le x \le 1.6$.

SOLUTION

a) To graph $t_1(x)$, we press: [Y=] [2nd] [DISTR] [4] to obtain **tpdf(**. We then enter the variable X and the number of degrees of freedom, 1, as shown in the top row of the display, and press [ENTER] .

To graph $t_3(x)$ and $t_{30}(x)$, we repeat the process above and define Y_2 and Y_3 as shown.

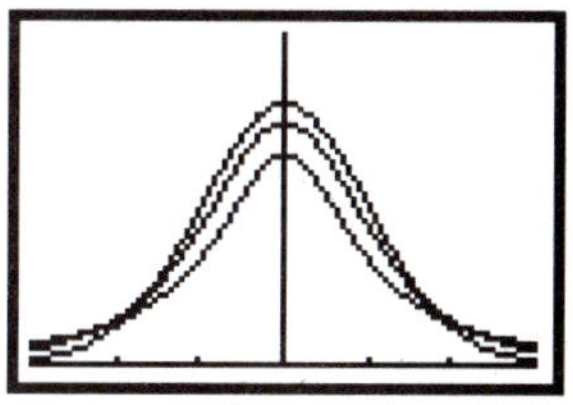

Upon setting the window to: $-3 \le x \le 3$; $0 \le y \le 0.5$, and pressing [GRAPH] , we obtain the graphs shown in the display.

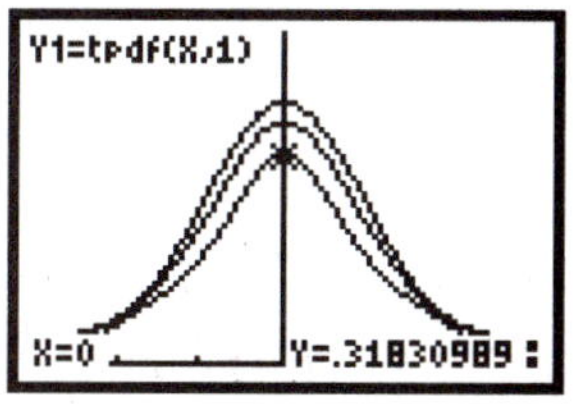

If we substitute $x = 0$ into the expression for $t_n(x)$, we find $t_n(0) = C_n$. To evaluate C_1 we press [TRACE] and obtain the display on the right showing that $C_1 = 0.318\ldots$
By tracing along the other curves, we find $C_3 = 0.367\ldots$ and $C_{30} = 0.395\ldots$
In Exercise 8, you will verify these values of C_n to 3 decimal digits.

b) Your TI-83 has a command **Shade_t(** which is analogous to the **ShadeNorm(** command for normal distributions. To access the **Shade_t(** command, press:

[2nd] [DISTR] [▶] [2]

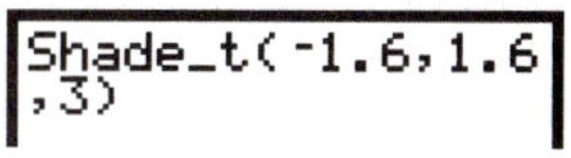

To shade the area under the graph of $t_3(x)$ and in the region $-1.6 \le x \le 1.6$, we enter in order, separated by commas, the lower bound -1.6, the upper bound 1.6, and the number of degrees of freedom, 3 as shown in the display. Upon pressing [ENTER], we obtain the shaded region shown in the display, indicating an area of 0.792….
That is, the probability that a random variable having a t distribution with 3 degrees of freedom takes a value between -1.6 and 1.6 is about 0.79.

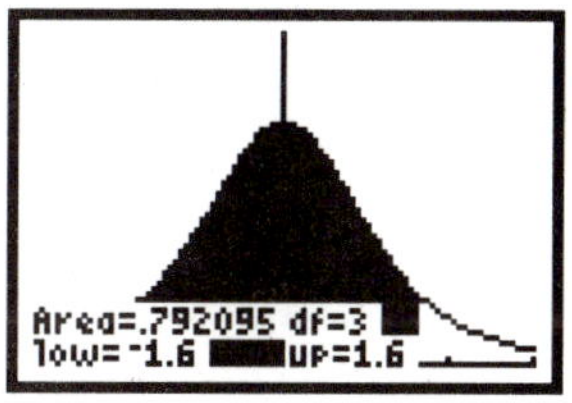

As we might expect, the solutions in *Worked Examples* 1 and 2 are in agreement. The TI-83 commands, **tpdf(** and **Shade_t(** enable you to calculate the t distribution probability density at any given value of x and the cumulative probability over any interval. When it is more convenient to calculate cumulative probabilities of the t distribution without sketching a graph, use the t distribution cumulative density function **tcdf(.** This command is found on the same menu as **tpdf(** and takes as input: lower bound, upper bound, and degrees of freedom. The **tcdf(** command can be used to graph the cumulative probability function and to calculate percentiles. (See Investigation 9 on page 56.)

Copyright © 1998 by Brendan Kelly Publishing Inc.

Exercises

1. Write the equation of the probability density function for the t distributions, $t_5(x)$, $t_{20}(x)$, and $t_{50}(x)$, using C_5, C_{20}, and C_{50} to denote the respective constants.

2. Graph each of the functions $t_n(x)/C_n$ in Exercise 1 in the range: $-3 \le x \le 3$; $0 \le y \le 1.5$.

a) Use your graphs to evaluate C_5, C_{20}, and C_{50}. Express the value of each constant to three decimal digits.

b) How does the graph of the t distribution change as the number of degrees of freedom is increased?

3. What is the probability that a variable which has a t distribution of 8 degrees of freedom will take a negative value?

4. a) Prove algebraically that $t_n(x) = t_n(-x)$ for all x.

b) What does the relationship in part a) tell us about the shape of the graph of $t_n(x)$?

c) Use the relationship in a) to determine the mean value of $t_n(x)$.

5. Shade the region under the graph of each t distribution for the given interval and state the area of the shaded portion.

a) $t_5(x)$: $-1 \le x \le 1$ b) $t_8(x)$: $-0.5 \le x \le 1$

c) $t_{10}(x)$: $x \le 1$ d) $t_{20}(x)$: $x > 0.8$

6. What is the probability that a variable with a t distribution of 10 degrees of freedom will assume a value less than 1.8?

7. a) What is the 50th percentile of the t distribution with 13 degrees of freedom?

b) Use the command **tcpf(** and a trial-and-error procedure to approximate the 60th and 95th percentiles of the t distribution with 13 degrees of freedom.

8. The exact value of the constant C_n in the definition of $t_n(x)$ is given by the following expressions.

$$C_n = \begin{cases} \dfrac{\sqrt{n}}{2^n}\ {}_{n-1}C_{\frac{n}{2}} & \text{for } n \text{ even} \\[2ex] \dfrac{2^{n-1}}{\sqrt{n}\,\pi\ {}_{n-1}C_{\frac{n-1}{2}}} & \text{for } n \text{ odd} \end{cases} \qquad \text{where } {}_nC_r = \frac{n!}{(n-r)!\,r!}$$

Do not confuse ${}_nC_r$ with the constant C_n.

a) Use this definition of C_n to obtain exact expressions for C_1, C_3, and C_{30}.

> Note: To evaluate ${}_nC_r$ on your TI-83, enter n and press MATH ◄ 3 . Then enter r.

b) Express C_1, C_3, and C_{30} in decimal form and compare with the values found in *Worked Example* 2.

c) Use the exact value of C_3 to construct the equation for $t_3(x)$. Verify that this yields the same graph as **tpdf(X,3)**.

Investigations

How Can You Find Percentiles for a *t* Distribution?

9. In *Worked Example* 3 of *Exploration* 12, you saw how the **invNorm(** command can be used to find any percentile for a normal distribution. Since the TI-83 does not have such an explicit inverse for the t distribution, you will need to use the cumulative probability distribution function **tcdf(** to determine t distribution percentiles. This investigation will help you discover how.

a) To find any percentile of a t distribution with 20 degrees of freedom, define **Y₁ = tcdf(-1E99,X,20)** as shown in the display below.

```
Plot1 Plot2 Plot3
\Y1■tcdf(-1E99,X
,20)
\Y2=
```

b) Graph **Y₁** in the window $-3 \le x \le 3$; $0 \le y \le 1$. Your graph should look like the one in the display.

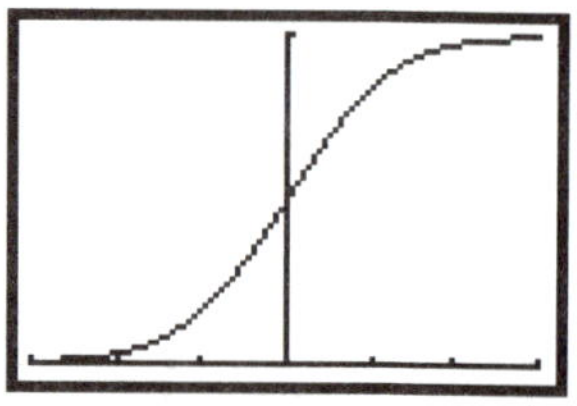

Explain the meaning of **Y₁**(x_0) for a given value x_0.

c) To find the 80th percentile of the t distribution with 20 degrees of freedom, graph **Y₂** = 0.8 on the same screen as **Y₁**.

d) Solve for the intersection of these two graphs by pressing:

2nd [CALC 5

Then press ENTER in response to each of the prompts, **First curve?**, **Second curve?**, and **Guess?**. Interpret the coordinates of the point of intersection shown on the display.

e) Give the 30th, 60th, 90th, and 95th percentiles of the t distribution with 20 degrees of freedom. Check your answers using the **tcpf(** command.

For Calculus Buffs

If you know calculus and you have mastered your limits, try these!

- Integrate the probability density function for $t_1(x)$ to find the precise value of C_1.
- Prove that the t distribution $t_n(x)$ approaches the normal distribution as $n \to \infty$.
- Integrate $t_3(x)$ to express the corresponding cumulative distribution function as a function of x.

Copyright © 1998 by Brendan Kelly Publishing Inc.

EXPLORATION 14 The Chi-Square Distribution

Karl Pearson
1857-1936

Karl Pearson was one of the giants in the development of modern statistics. Abandoning his early interest in law, he became a professor of applied mathematics at University College in London, England. His early work dealt with various topics including a treatment of the geometry of motion that anticipated some of the concepts presented in Einstein's *Theory of Relativity*. These developments were published in his book *The Grammar of Science* in 1892. Following in the tradition of Galton, Pearson studied evolution and heredity and between 1893 and 1912 published his statistical investigations in a series of 18 papers titled *Mathematical Contributions to the Theory of Evolution.* These contain developments of regression analysis, correlation coefficients, and the chi-square distribution.

The χ^2-distribution of n degrees of freedom has probability density function $\chi^2_n(x)$ given by:

$$\chi_n^{\,2}(x) = \begin{cases} K_n x^{\frac{n-2}{2}} e^{-\frac{x}{2}} & \text{for } x > 0 \leftarrow \text{where } K_n \text{ is a constant such that the area under the graph of the probability density function is 1.} \\ 0 & \text{elsewhere} \end{cases}$$

WORKED EXAMPLE 1

a) Find the values of the constant K_n for the χ^2_n distributions of $n = 1$, 4, and 5 degrees of freedom.

b) Graph $\chi^2_1(x)$, $\chi^2_4(x)$, and $\chi^2_5(x)$ in the window $0 \le x \le 10;\ 0 \le y \le 0.2$.

c) Determine the area under the graph of $\chi^2_5(x)$ and in the interval $2 \le x \le 6$, and interpret what this means.

SOLUTION

a) Substitution of $n = 1$ into the expression for χ^2_n yields $\chi^2_1(x) = \dfrac{K_1 e^{-(x/2)}}{\sqrt{x}}$.

To find the value of C_1, we graph $Y_1(x) = \dfrac{e^{-(x/2)}}{\sqrt{x}}$ in the window $0 \le x \le 10;\ 0 \le y \le 0.2$.

Then press 2nd [CALC] 7 and enter 0 and 10 in response to the prompts **Lower Limit?** and **Upper Limit?** followed by ENTER. We obtain the display shown here, indicating that the area under the curve is about 2.502.... To make the area under the graph of $\chi^2_1(x)$ equal to 1, we must divide by 2.502; i.e. set $K_1 = 2.502^{-1} \approx 0.4$.
Proceeding as above (with $0 < x \le 30$ and upper limit 30) we find $K_4 \approx 0.2500$ and $K_5 \approx 0.133$.

The area under the graph of $Y_1(x)$ in $0 < x \le 10$

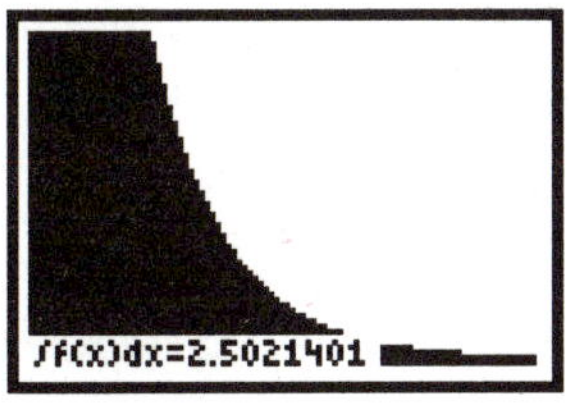

b) Substituting the values of K_n obtained in part a), we obtain:

$$\chi_1^2(x) \approx \frac{0.4e^{-(x/2)}}{\sqrt{x}}, \quad \chi_4^2(x) \approx \frac{1}{4} x e^{-x/2} \quad \text{and} \quad \chi_5^2(x) \approx 0.133 x^{3/2} e^{-x/2}$$

Entering these functions into the Y= editor and graphing yields the display on the right.

The $\chi^2_n(x)$ distributions with 1, 4 and 5 degrees of freedom

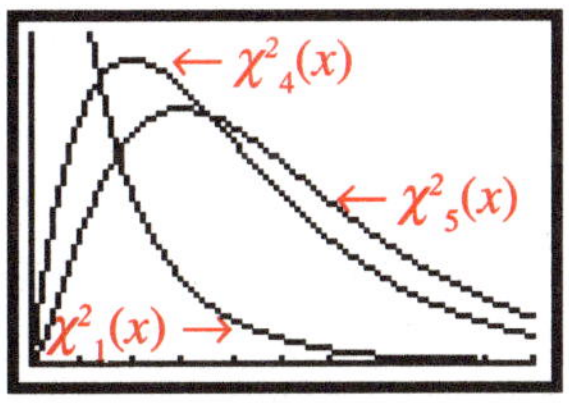

c) To determine the area under the graph of $\chi^2_5(x)$ in the interval $2 \le x \le 6$, we press:

2nd [CALC] 7 and enter 2 and 6 in response to the prompts **Lower Limit?** and **Upper Limit?**. Then we press ENTER and obtain the display shown here.
We see that the contained area is 0.543. That is, the probability that a random variable having a *chi-square* distribution with 5 degrees of freedom takes a value between 2 and 6 is about 0.54.

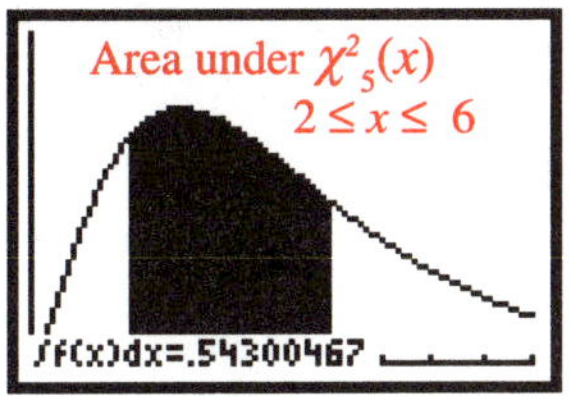

Copyright © 1998 by Brendan Kelly Publishing Inc.

In *Exploration* 13, you discovered that your TI-83 has the definition of the t distribution built into it. As you probably guessed, your TI-83 also has embedded in it the definition of the χ^2_n distribution. In the example that follows, we show how to use the TI-83 commands **χ²pdf(** (the χ^2_n probability density function) and **Shadeχ²(** to solve *Worked Example* 1.

Worked Example 2

a) Graph $\chi^2_1(x)$, $\chi^2_4(x)$ and $\chi^2_5(x)$ in the window $0 \le x \le 10$; $0 \le y \le 0.2$. Evaluate K_1, K_4, and K_5.

b) Determine the area under the graph of $\chi^2_5(x)$ and in the interval $2 \le x \le 6$.

Solution

a) To graph $\chi^2_1(x)$, we press: Y= 2nd [DISTR] 6 to obtain **χ²pdf(.** We then enter the variable X and the number of degrees of freedom, 1, as shown in the top row of the display, and press ENTER.

To graph $\chi^2_4(x)$ and $\chi^2_5(x)$, we repeat the process above and define Y_2 and Y_3 as shown.

Upon setting the window to: $0 \le x \le 10$; $0 \le y \le 0.2$, and pressing GRAPH, we obtain the graphs shown in the display.

If we substitute $x = 1$ into the expression for $\chi^2_n(x)$, we find $\chi^2_n(1) = K_n e^{-(1/2)}$. To evaluate K_1 we press 2nd [CALC] ENTER 1 ENTER and obtain the display on the right showing that $K_1 e^{-(1/2)} \approx 0.2419$, from which we calculate $K_1 \approx 0.398$ The same procedure applied to $\chi^2_4(x)$ and $\chi^2_5(x)$ yields $K_4 \approx 0.25$ and $K_5 \approx 0.133$. In Exercise 8, you will verify these values of K_n to 3 decimal digits.

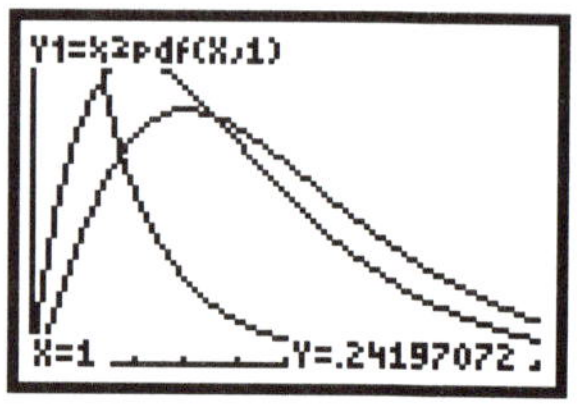

b) Your TI-83 has a command **Shadeχ²(** which is analogous to the **Shade_t(** command for the t distribution. To access the **Shadeχ²(** command, press:

2nd [DISTR] ▶ 3

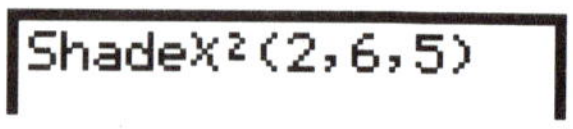

To shade the area under the graph of $\chi^2_5(x)$ and in the region $2 \le x \le 6$, we enter in order, separated by commas, the lower bound 2, the upper bound 6, and the number of degrees of freedom 5, as shown in the display. Upon pressing ENTER, we obtain the shaded region shown in the display, indicating an area of about 0.543.
That is, the probability that a random variable having a chi-square distribution with 5 degrees of freedom takes a value between 2 and 6 is about 0.54.

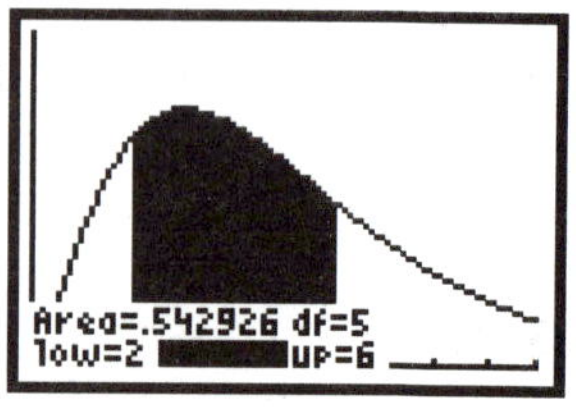

As we might expect, the solutions in *Worked Examples* 1 and 2 are in agreement. The TI-83 commands **χ²pdf(** and **Shadeχ²(** enable you to calculate the chi-square probability density at any given value of x and the cumulative probability over any interval. When it is more convenient to calculate cumulative probabilities of the chi-square distribution without sketching a graph, use the chi-square cumulative density function **χ²cdf(** This command is found on the same menu as **χ²pdf(** and takes as input: lower bound, upper bound, and degrees of freedom. The **χ²cdf(** command can be used to graph the chi-square cumulative probability function and to calculate percentiles. (See Investigation 9 on page 59.)

Copyright © 1998 by Brendan Kelly Publishing Inc.

Exercises

1. Write the equation of the probability density function for the chi-square distributions, $\chi^2_7(x)$, $\chi^2_9(x)$, and $\chi^2_{10}(x)$, using K_7, K_9, and K_{10} to denote the respective constants.

2. Graph each of the equations $\chi^2_n(x)/K_n$ in Exercise 1 in the range: $0 \le x \le 30$; $0 \le y \le 80$.

a) Use your graphs to evaluate K_7, K_9, and K_{10}. Express the value of each constant to three decimal digits.

b) How does the graph of the chi-square distribution change as the number of degrees of freedom is increased?

3. A variable X has a chi-square distribution with 8 degrees of freedom. Find the value of X associated with the highest probability density. (This is the *mode* of the distribution.)

4. a) The median of a continuous distribution is the 50th percentile. Use the command **χ²cdf(** and a tial-and-error procedure to estimate the median of χ^2_{13}, χ^2_{15}, χ^2_{25}, and χ^2_{30}.

b) Compare the median of each distribution in part a) with the corresponding number of degrees of freedom. Write a sentence to describe any relationship you discover.

c) Use the relationship in b) to estimate the median of χ^2_{40}.

5. Shade the region under the graph of each χ^2_n distribution for the given interval and state the area of the shaded portion.

a) χ^2_6: $0 \le x \le 5$ b) χ^2_{19}: $10 \le x \le 18$

c) χ^2_{11}: $x \le 7$ d) χ^2_{16}: $x \ge 10$

6. What is the probability that a variable with a χ^2 distribution of 10 degrees of freedom will assume a value less than 9?

7. Use the command **χ²cdf(** and a trial-and-error procedure to approximate the 60th and 95th percentiles of the χ^2 distribution with 13 degrees of freedom.

8. The exact value of the constant K_n in the definition of $\chi^2_n(x)$ is given by the following expressions.

$$K_n = \begin{cases} \dfrac{1}{2^{n/2}((n-2)/2)!} & \text{for } n \text{ even} \\[2ex] \dfrac{2^{(n-2)/2}((n-1)/2)!}{\sqrt{\pi}\,(n-1)!} & \text{for } n \text{ odd} \end{cases}$$

where ! is the factorial notation

a) Use this definition of K_n to obtain exact expressions for K_1, K_4, and K_5.

Note: To evaluate $n!$ on your TI-83, enter n and press MATH ◄ 4. Then press ENTER.

b) Express K_1, K_4, and K_5 in decimal form and compare with the values found in *Worked Example* 2.

c) Use the exact value of K_5 to construct the equation for $\chi^2_5(x)$. Verify that this yields the same graph as **χ²pdf(X,5).**

Investigations

How Can You Find Percentiles for a Chi-Square Distribution?

9. In *Investigation* 9 of *Exploration* 13, you discovered how to use the cumulative probability distribution function **tcdf(** to determine t distribution percentiles. In this investigation, you will follow the same procedure to determine percentiles for the chi-square distribution.

a) To find any percentile of a χ^2 distribution with 9 degrees of freedom, define **Y₁= χ²cdf(0,X,9)** as shown in the display below.

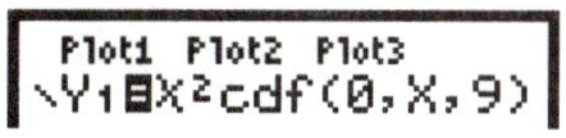

b) Graph **Y₁** in the window $0 \le x \le 25$; $0 \le y \le 1$. Your graph should look like the one in the display.

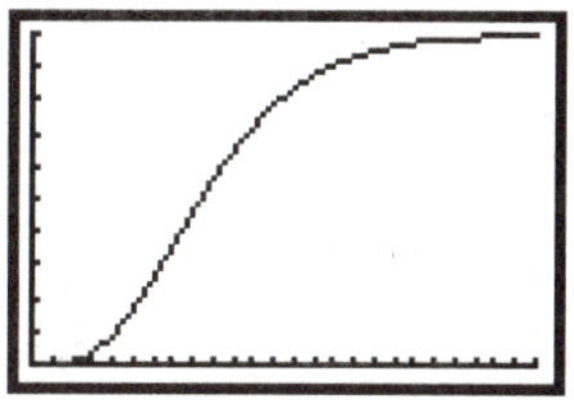

Explain the meaning of **Y₁**(x_0) for a given value x_0.

c) To find the 50th percentile of the χ^2 distribution with 9 degrees of freedom, graph **Y₂** = 0.5 on the same screen as **Y₁**.

d) Solve for the intersection of these two graphs by pressing:

2nd [CALC] 5

Then press ENTER in response to each of the prompts, **First curve?**, **Second curve?**, and **Guess?**. Interpret the coordinates of the point of intersection shown on the display.

e) Give the 30th, 60th, 90th, and 95th percentiles of the χ^2 distribution with 9 degrees of freedom. Check your answers using the **χ²cdf(** command.

For Calculus Buffs

If you know calculus and you like a challenge, try this.

The *gamma probability density function* $f(x)$ is defined by:

$f(x) = kx^{\alpha-1}e^{-x/\beta}$ for $0 < x < \infty$ and 0 for $x \le 0$.

where k is a constant such that the area under the graph of $f(x)$ is 1.

a) Substitute $t = x/\beta$ to show $\int_0^\infty kx^{\alpha-1}e^{-x/\beta}dx = k\beta^{\alpha-1}\Gamma(\alpha)$ where $\Gamma(\alpha) = \int_0^\infty t^{\alpha-1}e^{-t}dt$.

b) Express k in terms of α, β, and $\Gamma(\alpha)$.

c) Show that the gamma distribution with $\alpha = n/2$ and $\beta = 2$ is the chi-square distribution with n degrees of freedom.

d) Integrate by parts to show that $\Gamma(\alpha) = (\alpha-1)\Gamma(\alpha-1)$. Show that $\Gamma(\alpha) = (\alpha-1)!$ for integral α. Verify the formulas in Exercise 8.

Copyright © 1998 by Brendan Kelly Publishing Inc.

EXPLORATION 15 *The F Distribution*

CORBIS-BETTMANN

Sir Ronald A. Fisher
1890-1962

Sir Ronald A. Fisher, usually referred to as *R. A. Fisher*, was (along with Pearson and others) one of the founders of modern statistics. He graduated in 1912 from Cambridge with a degree in astronomy. Attempts to quantify the errors inherent in astronomical measurements drew him into the exploration of statistical concepts, which he subsequently applied to the study of evolution and the genetics of human blood groups. Fisher followed Gossett's tradition in developing the theory of distributions pertaining to small samples. His main contribution was in the discovery of a technique, known as the *analysis of variance*, that is widely used in the social sciences to identify the importance of each contributary cause to a final outcome. The technique involves a test of significance based upon the properties of a distribution that is called the *F distribution* in his honor.

The F distribution of n_1 and n_2 degrees of freedom has probability density function $f_{n_1,n_2}(x)$ given by:

$$f_{n_1,n_2}(x) = \begin{cases} \dfrac{k}{\left(1+\dfrac{n_1}{n_2}x\right)^{(n_1+n_2)/2}} x^{(n_1/2)-1} & \text{for } x > 0 \\ 0 & \text{elsewhere} \end{cases}$$

← where k is a constant such that the area under the graph of the probability density function is 1.

Worked Example 1

a) Find the value of the constant k for the F distribution of 5 and 20 degrees of freedom.

b) Graph the distribution in part a) in the window $0 \le x \le 5$; $0 \le y \le 1$.

c) Determine the area under the graph of $f_{5,20}$ and in the interval $3 < x < \infty$, and interpret what this means.

Solution

a) Substitution of $n_1 = 5$, $n_2 = 20$ into the expression for $f_{5,20}$ yields $f_{5,20}(x) = \dfrac{kx^{3/2}}{\left(1+\frac{x}{4}\right)^{25/2}}$.

To find the value of k, we graph $Y_1(x) = \dfrac{x^{3/2}}{\left(1+\frac{x}{4}\right)^{25/2}}$ in the window $0 \le x \le 5$; $0 \le y \le 0.1$.

Then press **2nd** [CALC] **7** and enter 0 and 5 in response to the prompts **Lower Limit?** and **Upper Limit?** followed by **ENTER**. We obtain the display shown here, indicating that the area under the curve is about 0.1123. To make the area under the graph of $f_{5,20}$ equal to 1, we must divide by 0.1123; i.e. set $k = 0.1123^{-1} \approx 8.90$.

The area under the graph of Y_1 in $0 < x \le 5$

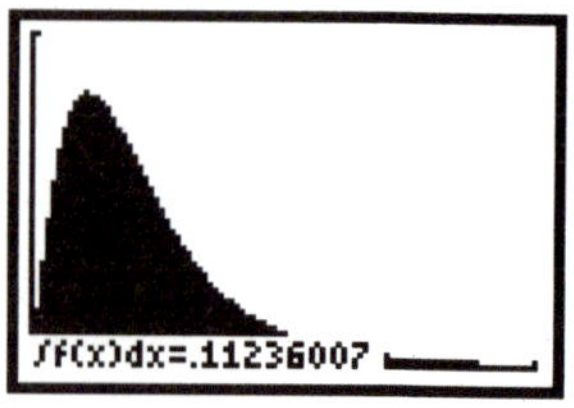

b) Substitution of the value of k obtained in part a), yields: $f_{5,20}(x) = 8.90\dfrac{x^{3/2}}{\left(1+\frac{x}{4}\right)^{25/2}}$

Entering this function into the Y= editor and graphing yields the display on the right.

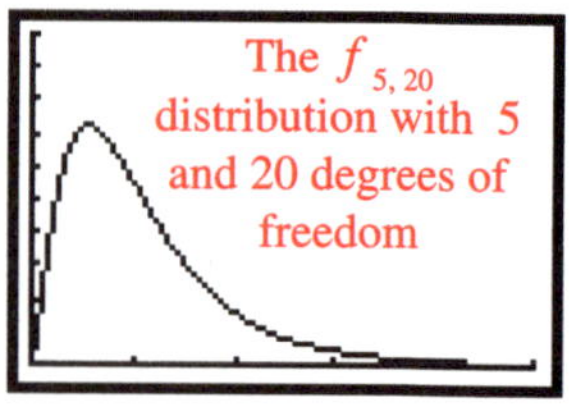

c) To determine the area under the graph of $f_{5,20}(x)$ in the interval $3 < x < \infty$, we press:

2nd [CALC] **7** and enter 0 and 3 in response to the prompts, **Lower Limit?** and **Upper Limit?**. Then we press **ENTER** and obtain the display shown here.
We see that the contained area is about 0.968. That is, the probability that a random variable having an F distribution with 5 and 20 degrees of freedom takes a value between 0 and 3 is about 0.968. Hence the probability that this distribution takes a value greater than 3 is $1 - 0.968$ or about 0.032. That is, a variable with an $f_{5,20}$ distribution takes a value greater than 3 only about 3.2% of the time.

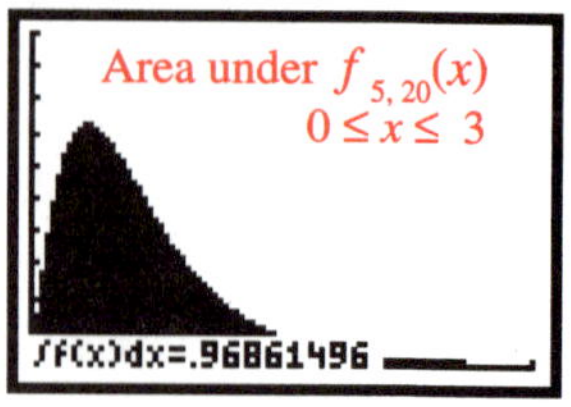

Copyright © 1998 by Brendan Kelly Publishing Inc.

Your TI-83 has embedded in it also the definition of the *F* distribution. In the example that follows, we show how to use the TI-83 commands **Fpdf(** (the *F* probability density function) and **ShadeF(** to solve *Worked Example* 1.

Worked Example 2

a) Graph $f_{5,\,20}(x)$ in the window $0 \le x \le 5; 0 \le y \le 1$. Evaluate *k*.

b) Determine the area under the graph of $f_{5,\,20}(x)$ and in the interval $3 < x < \infty$.

c) Determine the 95th percentile of the *F* distribution of 5 and 20 degrees of freedom. Explain the meaning of the 95th percentile of this distribution.

Solution

a) To graph $f_{5,\,20}(x)$, we press: Y= 2nd [DISTR] 8 to obtain **Fpdf(**. We then enter the variable X and the number of degrees of freedom, n_1, n_2 as shown in the display, and press ENTER.

Upon setting the window to: $0 \le x \le 5; 0 \le y \le 1$, and pressing GRAPH, we obtain the graph shown in the display.

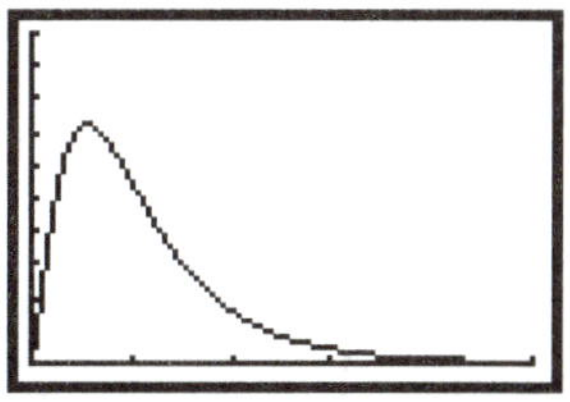

If we substitute $x = 1$ into the expression for $f_{5,\,20}(x)$, we find $f_{5,\,20}(1) = (4/5)^{25/2}k$. To evaluate *k* we press 2nd [CALC] ENTER 1 ENTER and obtain the display on the right showing that $(4/5)^{25/2}k \approx 0.5449$. From this we calculate $k \approx 8.865$. In the ***Challenge*** on page 62, you will verify this value of *k* to 3 decimal digits.

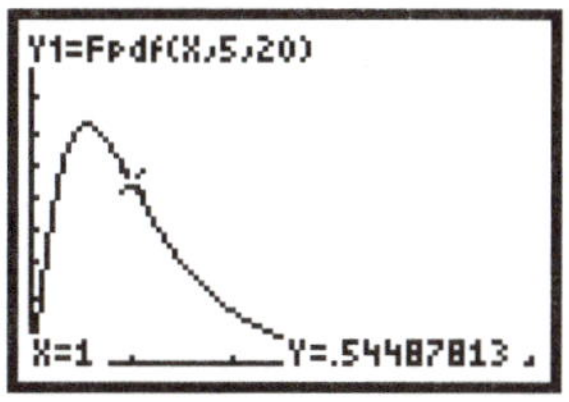

b) Your TI-83 has a command **ShadeF(** which is analogous to the **Shade_t(** command for the *t* distribution. To access the **ShadeF(** command, press:

2nd [DISTR] ▶ 4

To shade the area under the graph of $f_{5,\,20}(x)$ and in the region $0 \le x \le 3$, we enter in order, separated by commas, the lower bound 0, the upper bound 3, and the number of degrees of freedom, 5 and 20, as shown in the display. Upon pressing ENTER, we obtain the shaded region shown in the display, indicating an area of about 0.965. Hence the area in the region $3 < x < \infty$ is about $1 - 0.965 \approx 0.035$.

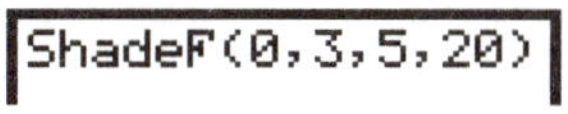

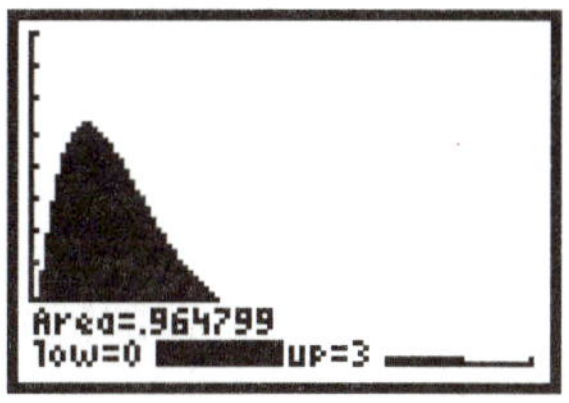

c) The 95th percentile of the *F* distribution $f_{5,\,20}(x)$ is the value of *x*, denoted $F_{.95}$ such that 95% of the area under this probability density function lies to the left of $x = F_{.95}$. That is, **ShadeF(0, $F_{.95}$, 5, 20)** yields an area of 0.95. We could determine $F_{.95}$ by calculating the cumulative density function **Fcdf(0, x_0, 5, 20)** for various values of x_0, but that process is a hit-and-miss approach.

Alternatively, we observe that the 95th percentile is the value of *x* at which the cumulative density function takes the value 0.95. Therefore, we graph the cumulative density function, by defining Y_1 = **Fcdf(0, X, 5, 20)** and $Y_2 = 0.95$, as in the display. We then press GRAPH and solve for the intersection of these two graphs by pressing:

2nd [CALC] 5

We press ENTER in response to each of the prompts, **First curve?**, **Second curve?**, and **Guess?**. The display shows that the 95th percentile is $F_{.95} \approx 2.71$. That is, the probability that the *F* distribution with 5 and 20 degrees of freedom takes a value ≤ 2.71 is 0.95. Conversely, the probability that it takes a value greater than 2.71 is only 0.05.

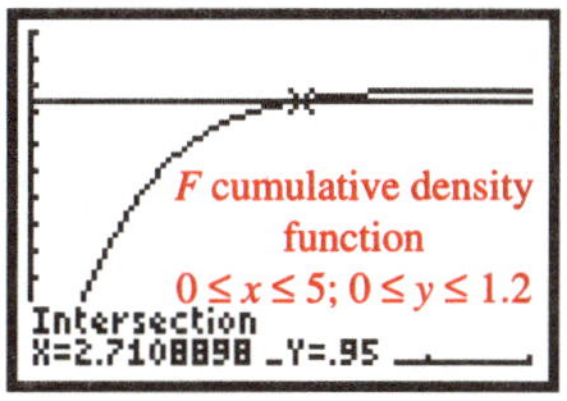

Copyright © 1998 by Brendan Kelly Publishing Inc.

Exercises

1. Write the equation of the probability density function for the F distribution with the given numbers of degrees of freedom. Use k to denote the constant.

a) $n_1 = 4, n_2 = 6$ b) $n_1 = 5, n_2 = 6$ c) $n_1 = 5, n_2 = 7$

2. Graph each of the functions $f_{n_1,n_2}(x)/k$ in exercise 1 in the range: $0 \le x \le 5$; $0 \le y \le 0.15$.

a) Use your graphs to evaluate k in each case. Express the value of each constant to three decimal digits.

b) How does the graph of the F distribution change as n_1 is increased? as n_2 is increased?

3. A variable X has an F distribution with $n_1 = 8$, $n_2 = 10$ degrees of freedom. Find the value of X associated with its highest probability density. (This is the *mode.*)

4. Shade the region under the graph of the probability density function of each F distribution for the given interval and state the area of the shaded portion.

a) $f_{6,14}$: $0 \le x \le 1.4$ b) $f_{10,16}$: $1 \le x \le 1.6$

c) $f_{5,3}$: $x \le 0.5$ d) $f_{9,18}$: $x \ge 1.5$

5. What is the probability that a variable with an F distribution of 10 and 13 degrees of freedom will take a value less than 2?

6. The display shows the graph of the probability density function $f_X(x)$ in the window: $0 \le x \le 3$; $0 \le y \le 1$.

Write an expression in terms of $F_X(x)$, the cumulative distribution function, for the area under the graph that is contained in each interval.

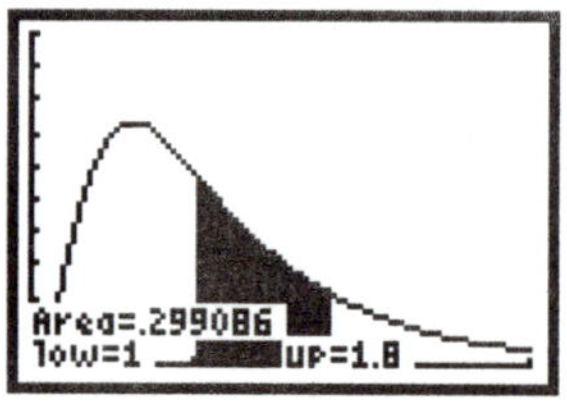

a) $0 \le x \le 1$

b) $0 \le x \le 1.8$

c) $1 \le x \le 1.8$

d) $0 \le x \le \infty$

e) $1.8 \le x \le \infty$

f) For which of these intervals can you find a numerical value of the area?

7. The function graphed in Exercise 6 is the probability density function of the F distribution with 7 and 12 degrees of freedom.

a) Use your TI-83 to graph the corresponding cumulative distribution function.

b) Use this graph to verify the area given in Exercise 6.

c) Use the command **Fpdf(** to verify your answer in part b).

8. a) Investigation 9 of *Explorations* 13 and 14 show how to find percentiles for the t and χ^2 distributions. Follow a similar procedure to determine the 90th and the 95th percentiles of an F distribution with 17 and 14 degrees of freedom.

b) Find the values α and β such that $F_{17,14}(\alpha) = 0.90$ and $F_{17,14}(\beta) = 0.95$ where $F_{17,14}(x)$ is the cumulative distribution function in part a). What is the probability that a value assumed by this F distribution lies in the interval $\alpha \le x \le \beta$?

Investigations

9. In Exercise 8, you used the techniques introduced in *Explorations* 13 and 14 to find any specified percentile of the F distribution. This enabled you to find the interval that bounds the "middle" 90% of the area under the probability density function. The unshaded area on the left contains all the values of x such that $F(x) \le 0.05$. The unshaded area on the right contains all the values of x such that $F(x) > 0.95$. We can say that 90% of the time, the F distribution will assume a value that is no less than α and no greater than β. For this reason, $\alpha \le x \le \beta$ is called a *90% confidence interval.*

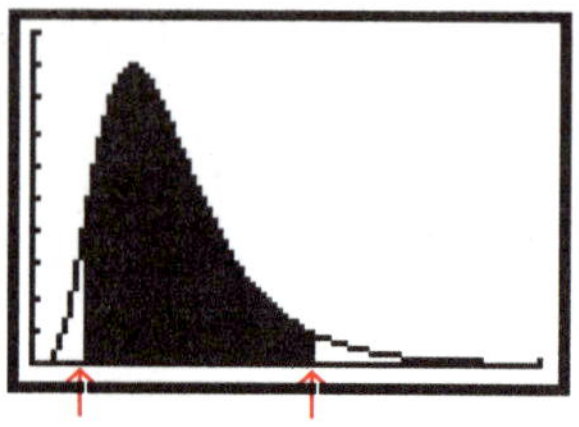

$x = \alpha$ $x = \beta$

$F_{.05,14,20}$ $F_{.95,14,20}$

Area within $\alpha \le x \le \beta$ is 0.9.

In this investigation you discover how to find the "middle" 90% confidence interval for *any* distribution.

a) What percentile is α? What percentile is β?

b) The displays above show the probability density function for the F distribution with 14 and 20 degrees of freedom. The 90% confidence interval is $F_{.05,14,20} < x < F_{.95,14,20}$. Explain the meanings of $F_{.05,14,20}$ and $F_{.95,14,20}$.

c) Graph the F distribution with 22 and 30 degrees of freedom and find the values of $F_{.05,22,30}$ and $F_{.95,22,30}$.

d) Write the 90% confidence interval for $F_{22,30}$.

e) Write the 90% confidence interval for $F_{30,22}$.

f) Take the reciprocals of $F_{.05,22,30}$ and $F_{.95,22,30}$ and compare with the values of $F_{.05,30,22}$ and $F_{.95,30,22}$. Describe any relationship you discover.

CHALLENGE

On page 60, we defined the probability density function of the F distribution of n_1 and n_2 degrees of freedom by:

$$f_{n_1,n_2}(x) = \frac{k}{\left(1+\frac{n_1}{n_2}x\right)^{(n_1+n_2)/2}} x^{(n_1/2)-1} \quad \text{for } x > 0$$

The constant k is given by $k = \dfrac{\Gamma\left(\frac{n_1+n_2}{2}\right)}{\Gamma\left(\frac{n_1}{2}\right)\Gamma\left(\frac{n_2}{2}\right)}\left(\dfrac{n_1}{n_2}\right)^{n_1/2}$

where $\Gamma(n) = (n-1)!$ for integral n.

Furthermore $\Gamma(x+1) = x\Gamma(x)$ for all real x.

Use the identities above and the fact that $\Gamma\left(\frac{1}{2}\right) = \frac{1}{\sqrt{\pi}}$ to calculate k for $f_{5,20}$. Compare with the value in *Worked Example* 2.

Copyright © 1998 by Brendan Kelly Publishing Inc.

EXPLORATION 16 Related Distributions

Suppose we sample values from a normally distributed variable X, and we square these values. What is the distribution of this new set of numbers? That is, what is the distribution of the random variable X^2? In general, if random variables X and Y are related by $Y = g(X)$, where g is a monotonically increasing or decreasing differentiable function, how can we compute the probability density function f_Y from the density function f_X and g? The following theorem (easily proved using elementary calculus), will enable you to answer this question and to discover some interesting relationships among the distributions you encountered in the earlier *Explorations.*

> **Theorem**
>
> Let X and Y denote continuous random variables with probability density functions $f_X(x)$ and $f_Y(y)$. If $y = g(x)$, and $\frac{dy}{dx} \neq 0$ throughout the range of values of x, then
>
> $$f_Y(y) = \left|\frac{dx}{dy}\right| f_X(x)$$

← *If you haven't studied calculus, don't worry.* $\left|\frac{dx}{dy}\right|$ *will be provided for you as needed.*

Worked Example 1

a) Suppose X is a normally distributed variable with mean 0 and standard deviation 1. Graph the cumulative distribution functions of X and the random variable Y, where Y = 3X in the window $-4 \le x \le 4$; $0 \le y \le 1$.

b) Graph the corresponding probability density functions in part a) in the window $-5 \le x \le 5$; $0 \le y \le 0.4$ and compare their means and standard deviations.

Solution

a) Let $F_X(x)$ and $F_Y(x)$ denote the cumulative distribution functions of X and Y. Then from the definition of a cumulative distribution function:

$$F_Y(x) = \text{Prob}(Y \le x) = \text{Prob}(3X \le x) = \text{Prob}(X \le x/3) = F_X(x/3)$$

To graph the cumulative distribution function $F_X(x)$ in the interval $-4 \le x \le 4$, we define **Y₁ = normalcdf(-4, X)** (accessible by pressing **2nd** [DISTR]) . To graph $F_Y(x)$ in the interval $-4 \le x \le 4$, we define **Y₂ = normalcdf(-4, X/3)** as shown in the display. Upon pressing **ENTER**, we obtain the graphs shown.

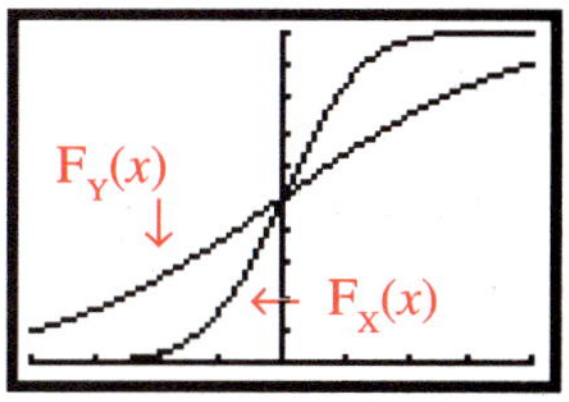

b) Since X has the standard normal distribution, its density function is given by $f_X(x) = \frac{1}{\sqrt{2\pi}} e^{-x^2/2}$ which is embedded in your TI-83 as **normalpdf(X)**.

To obtain $f_Y(x)$, we apply the theorem above to obtain $f_Y(3x) = \left|\frac{dx}{dy}\right| f_X(x) = \frac{1}{3} f_X(x)$, and so $f_Y(x) = \frac{1}{3} f_X(x/3)$; i.e. $f_Y(x) = \frac{1}{3\sqrt{2\pi}} e^{-x^2/18}$ which is stored in the TI-83 as **(1/3)normalpdf(X/3)**. To graph $f_X(x)$ and $f_Y(x)$, we could enter their equations into the Y= editor. However, it is more convenient to define Y_3 and Y_4 as shown in the display. Upon graphing these equations in the given window, we obtain this display. We observe that both graphs are symmetrical about the y-axis, so every value of x has the same probability density as $-x$, and the means of both distributions are 0. The standard deviation measures how close the distribution clusters around the mean. The graphs show that the standard deviation of $f_Y(x)$ is much greater than that of $f_X(x)$. In fact, inspection of the equation for $f_Y(x)$ given above, when compared with the equation given on page 50, reveals that the standard deviation is 3.

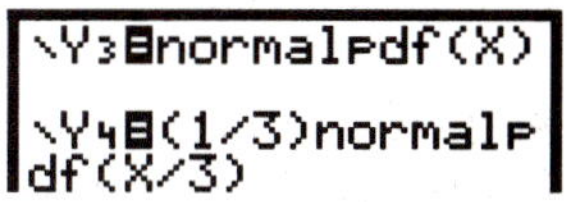

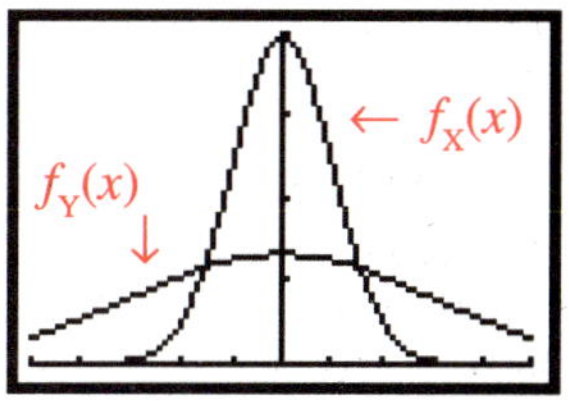

Copyright © 1998 by Brendan Kelly Publishing Inc.

Worked Example 2

a) Given a normally distributed random variable X with mean 0 and standard deviation 1, graph the cumulative distribution function of X and the cumulative distribution function of the random variable Y where $Y = X^2$.

b) Compute the probability density function $f_Y(x)$ and identify the distribution of Y. Graph the cumulative distribution of $\chi^2_1(x)$ in the window: $-4 \le x \le 4; 0 \le y \le 1$ to verify your computations.

Solution

a) Let $F_X(x)$ and $F_Y(x)$ denote the cumulative distribution functions of X and Y. Then from the definition of a cumulative distribution function, for $x \ge 0$:

$$F_Y(x) = \text{Prob}(Y \le x) = \text{Prob}(X^2 \le x) = \text{Prob}(-\sqrt{x} \le X \le \sqrt{x}) .$$

Since $f_X(x)$ is symmetric about the y-axis, i.e. $f_X(x) = f_X(-x)$,

$$\text{Prob}(-\sqrt{x} \le X \le \sqrt{x}) = 2F_X(\sqrt{x}) - 1 \qquad \text{(See Exercise 2)}$$

That is, $F_Y(x) = 2F_X(\sqrt{x}) - 1$ where $x \ge 0$. ①

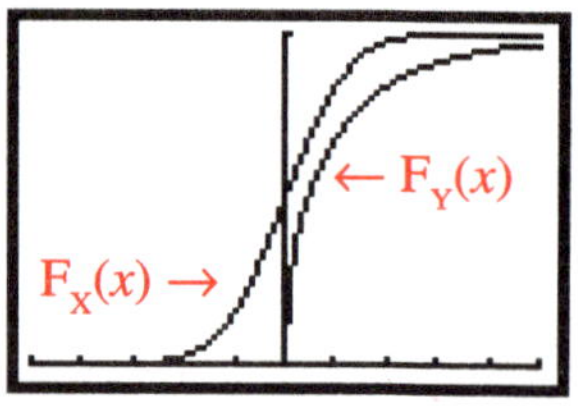

In *Worked Example* 1, we observed that the cumulative distribution function of X can be graphed on the TI-83 by defining **Y$_1$ = normalcdf(-4,X).** From equation ① above, it follows that we can graph the cumulative distribution function $F_Y(x)$ by defining Y_2 = **2[1/2 + normalcdf(0, √ X)] - 1 = 2normalcdf(0,√ X),** as shown in the display.

b) Since X has the standard normal distribution, its density function is given by

$$f_X(x) = \frac{1}{\sqrt{2\pi}} e^{-x^2/2}$$ which is embedded in your TI-83 as **normalpdf(X)**.

To obtain $f_Y(x)$, we apply the theorem on page 63 to obtain $f_Y(x^2) = \left|\frac{dx}{dy}\right| f_X(x)$. ← Here $y = x^2$, so by elementary calculus $\left|\frac{dx}{dy}\right| = \frac{1}{2x}$

Since $\left|\frac{dx}{dy}\right| = \frac{1}{2x}$, we write $f_Y(x^2) = \frac{1}{2x} f_X(x)$, so $f_Y(x) = \frac{1}{2\sqrt{x}} f_X(\sqrt{x})$; for $x \ge 0$.

i.e. $f_Y(x) = \frac{1}{2\sqrt{2\pi}} \bullet \frac{1}{\sqrt{x}} e^{-(\sqrt{x})^2/2}$ which simplifies to $f_Y(x) = \frac{1}{2\sqrt{2\pi}} x^{-1/2} e^{-x/2}$. That is:

$$f_{X^2}(x) = \begin{cases} \frac{1}{2\sqrt{2\pi}} x^{-1/2} e^{-x/2} & x \ge 0 \\ 0 & x < 0 \end{cases}$$

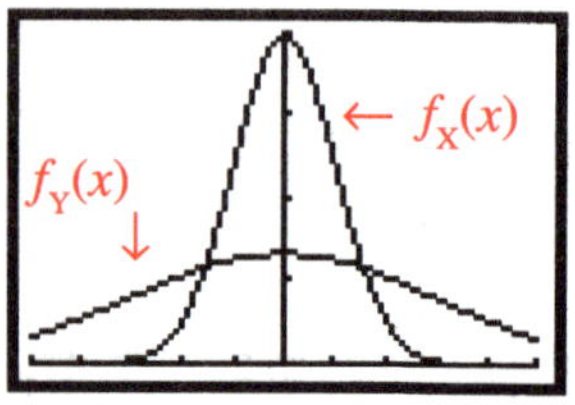

By comparing this probability density function with the density function $\chi_n^2(x)$ given on page 57, we observe that $f_{X^2}(x) = \chi_1^2(x)$. That is, X^2 has a chi-square distribution with one degree of freedom. To verify graphically that these distributions are the same, we could graph the cumulative distribution function of $\chi_1^2(x)$ by defining $Y_3(x) = \chi^2$**cdf(0,X,1)** and observing that the graphs of Y_2 and Y_3 are identical.

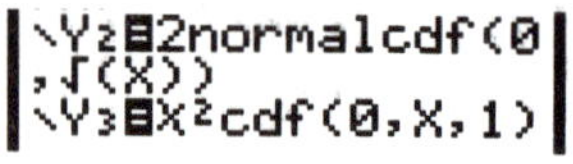

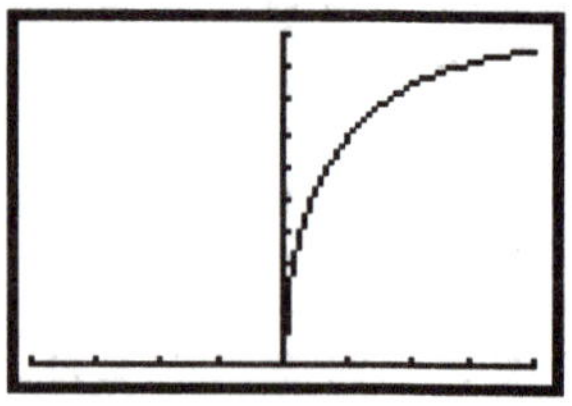

Graphs of cumulative distribution functions of X^2 and $\chi_1^2(x)$ are coincident.

In *Worked Example* 2, we proved that if X_1 is a normally distributed random variable with $\mu = 0$ and $\sigma = 1$, then the random variable X^2 has the chi-square distribution with one degree of freedom. It so happens that if X_1 and X_2 are two normally distributed random variables with $\mu = 0$ and $\sigma = 1$, then the random variable Y, where $Y = X_1^2 + X_2^2$, has a chi-square distribution with *two* degrees of freedom. The following theorem shows the general case.

> **Theorem:** If $X_1, X_2, \ldots X_n$ are independent normally distributed random variables with $\mu = 0$ and $\sigma = 1$, then random variable $Y = X_1^2 + X_2^2 + \ldots + X_n^2$ has the chi-square distribution with n degrees of freedom.

Copyright © 1998 by Brendan Kelly Publishing Inc.

Worked Example 3

a) If X denotes a random variable having an F distribution with n_1 and n_2 degrees of freedom, determine the distribution of the random variable Y = 1/X.

b) Use what you discover in part a) to prove $F_{1-\alpha,n_1,n_2} = \dfrac{1}{F_{\alpha,n_2,n_1}}$ where F_{α,n_2,n_1} is the αth percentile of F_{n_2,n_1}.

Solution

a) Using the *Change of Variable* theorem on page 63, we write: $f_{1/X}(1/x) = \left|\dfrac{dx}{dy}\right| f_X(x)$, ①

where $f_X(x)$ and $f_{1/X}(x)$ denote the probability density functions of X and 1/X respectively.

Furthermore $y = 1/x$, and therefore $\left|\dfrac{dx}{dy}\right| = x^2$. ②

Since X has the F distribution with n_1 and n_2 degrees of freedom,

$$f_X(x) = \frac{k}{\left(1+\frac{n_1}{n_2}x\right)^{(n_1+n_2)/2}} x^{(n_1/2)-1} \text{ for } x \ge 0 \text{ and 0 elsewhere.} \qquad ③$$

Substituting from ② and ③ into ① yields: $f_{1/X}(1/x) = x^2 \dfrac{k}{\left(1+\frac{n_1}{n_2}x\right)^{(n_1+n_2)/2}} x^{(n_1/2)-1}$ ④

Replacing $1/x$ by x in equation ④ and simplifying, we obtain (see Exercise 5):

$$f_{1/X}(x) = \frac{K}{\left(1+\frac{n_2}{n_1}x\right)^{(n_1+n_2)/2}} x^{(n_2/2)-1} \text{ where } K \text{ is the constant } K = \left(\frac{n_2}{n_1}\right)^{\frac{n_1+n_2}{2}} k. \qquad ⑤$$

By comparing ⑤ with the definition of the F distribution on page 60, we see that $f_{1/X}(x)$ defines a random variable having the F distribution with n_2 and n_1 degrees of freedom.

b) Let $F_{n_1,n_2}(x)$ and $F_{n_2,n_1}(x)$ denote the cumulative distribution functions of X and 1/X in a).

Since $\text{Prob}(X \le x) + \text{Prob}\left(\frac{1}{X} < \frac{1}{x}\right) = 1$ (see Exercise 6), then $F_{n_1,n_2}(x) + F_{n_2,n_1}(1/x) = 1$; that is, $F_{n_1,n_2}(x) = 1 - F_{n_2,n_1}(1/x)$. ⑥

Substituting F_{α,n_1,n_2} for x in equation ⑥ yields $F_{n_1,n_2}(F_{\alpha,n_1,n_2}) = 1 - F_{n_2,n_1}\left(\dfrac{1}{F_{\alpha,n_1,n_2}}\right)$ ⑦

By definition, the left side of equation ⑦ is α, so $F_{n_2,n_1}\left(\dfrac{1}{F_{\alpha,n_1,n_2}}\right) = 1 - \alpha$, as required to prove.

This *Exploration* shows some of the techniques mathematicians use to relate the distribution of one or more random variables to known distributions. Two of the most important relationships are the following:

- If $X_1, X_2, \ldots X_n$ are n normally distributed variables with mean μ and standard deviation σ, then the mean $\overline{X} = \dfrac{X_1 + X_2 + \ldots + X_n}{n}$ is normally distributed with mean μ and standard deviation $\sigma/\sqrt{n}$

- If X_1 and X_2 are random variables having chi-square distributions with n_1 and n_2 degrees of freedom, then $Y = \dfrac{X_1/n_1}{X_2/n_2}$ has the F distribution with n_1 and n_2 degrees of freedom.

Copyright © 1998 by Brendan Kelly Publishing Inc.

Exercises

1. A random variable X has a probability density function $f_X(x)$. To find the probability density function for the random variable $1/X$, a student substitutes $1/x$ for x and computes $f_X(1/x)$. Is this the correct probability density function? Explain why or why not.

2. The display shows the graph of a normally distributed random variable X with $\mu = 0$ and $\sigma = 1$. The shaded portion is contained in the interval $-\sqrt{x} \le X \le \sqrt{x}$.

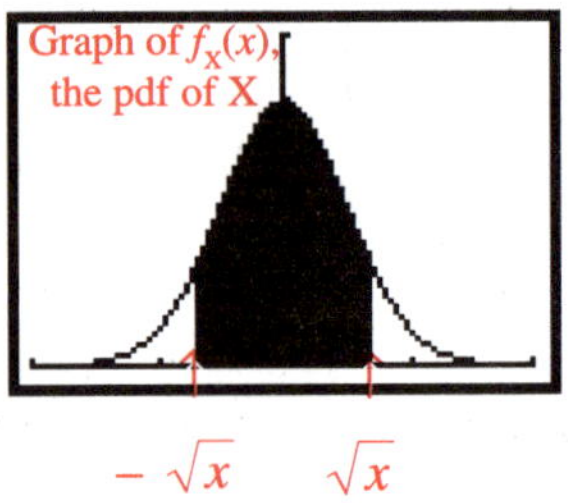

If $F_X(x)$ denotes the cumulative distribution function, express each area in terms of $F_X(x)$.

a) The unshaded area left of $x = -\sqrt{x}$ and under the graph.

b) The unshaded area right of $x = \sqrt{x}$ and under the graph.

c) The shaded area in the interval $-\sqrt{x} \le X \le \sqrt{x}$.

d) Prove that the areas in parts a) and b) are equal.

e) Prove equation ① in *Worked Example* 2; i.e.

$$F_Y(x) = 2F_X(\sqrt{x}) - 1$$

3. a) A random variable X has a t distribution with 8 degrees of freedom. Graph the cumulative distribution function of X^2.
b) Graph the cumulative distribution function $F_{1,8}(x)$. Explain what you discover.

4. a) If $f_X(x)$ denotes the probability density function of a random variable X, express the probability density function $f_Y(x)$ of the random variable $Y = 5X$ in terms of $f_X(x)$.

(Hint: If $y = 5x$, then $\left|\frac{dx}{dy}\right| = \frac{1}{5}$.)

b) If $f_X(0.5) = 0.04$, what is the value of $f_Y(2.5)$?

5. Prove that equation ⑤ in *Worked Example* 3 follows from equation ④.

6. Show that for any random variable X and any real number x where $x \ne 0$, $\text{Prob}\,(X \le x) + \text{Prob}\left(\frac{1}{X} < \frac{1}{x}\right) = 1$.

7. a) Graph the probability density function $f_{3,5}(x)$ of the F distribution with 3 and 5 degrees of freedom in the window $0 \le x \le 4$; $0 \le y \le 1$.
b) Graph $f_{3,5}(1/x)$ in the same window as in part a) and compare the graphs. For what value of x do the graphs intersect?

8. a) Graph $F_{3,5}$ to find its 95th percentile. Then graph $F_{5,3}$ to find its 5th percentile. Verify that $F_{.95,3,5} \times F_{.05,5,3} = 1$.
b) Find a 90% confidence interval for $F_{3,5}$.

Investigations

9. In Exercise 3, you may have discovered that if X is a random variable with a t distribution of 8 degrees of freedom, then X^2 has an F distribution with 1 and 8 degrees of freedom. In this investigation, you will follow the method of *Worked Example* 2b) to prove this relationship for any number of degrees of freedom; that is,

If X is a random variable with a t distribution of n degrees of freedom, then X^2 has an F distribution with 1 and n degrees of freedom.

a) Let $t_n(x)$ and $t_n^{\,2}(x)$ denote the probability density functions of X and X^2. Write an expression for $t_n^{\,2}(y)$ in terms of $t_n(x)$ where $y = x^2$.

b) Replace x with $\sqrt{y}$ in your expression in part a).

c) Replace y with x in the expression in part b). Why is it legitimate to do this?

d) Compare your probability density function in part c) with the probability density function $f_{1,n}(x)$ of the F distribution.

For Calculus Buffs

If you have a facility with calculus, you may want to tackle the proof of this very important theorem.

Theorem

If X is a normally distributed random variable with mean 0 and standard deviation 1, and Y is an independent random variable having a chi-square distribution with n degrees of freedom, then the variable Z where $Z = \dfrac{X}{\sqrt{Y/n}}$ has a t distribution with n degrees of freedom.

This proof generalizes the method in the *Worked Examples*.

a) Form the product of the probability density functions of X and Y/n to obtain the joint probability density function $f_{XY}(x, y)$

b) Differentiate $x = z\sqrt{y/n}$ with respect to z while holding y constant to obtain $\left|\frac{dx}{dz}\right|$.

c) Apply the generalized change of variable theorem:

$$f_{ZY}(z, y) = f_{XY}(x, y)\left|\frac{dx}{dz}\right|$$

to obtain the probability density function in y and z.

d) To integrate, substitute $u = \frac{y}{2}\left(1 + \frac{z^2}{n}\right)$, and integrate with respect to u from 0 to ∞. Compare the remaining integrand with the probability density function for $t_n(x)$.

Copyright © 1998 by Brendan Kelly Publishing Inc.

UNIT 5 HYPOTHESIS TESTING & CONFIDENCE INTERVALS

ONE SAMPLE

σ known

HYPOTHESIS $\mu = \mu_0$

Z-Test

INPUT
```
Z-Test
 Inpt:Data Stats
 μ0:2.4
 σ:2
 x̄:2.9
 n:80
 μ:≠μ0 <μ0 >μ0
 Calculate Draw
```

OUTPUT
```
Z-Test
 μ≠2.4
 z=2.236067977
 p=.0253472347
 x̄=2.9
 n=80
```

CONFIDENCE INTERVAL

INPUT
```
ZInterval
 Inpt:Data Stats
 σ:2
 x̄:2.9
 n:80
 C-Level:.95
 Calculate
```

OUTPUT
```
ZInterval
 (2.4617,3.3383)
 x̄=2.9
 n=80
```

This means that we can assert with a 95% degree of confidence that the interval (2.461… , 3.338…) contains the true mean.

σ unknown

HYPOTHESIS $\mu = \mu_0$

T-Test

INPUT
```
T-Test
 Inpt:Data Stats
 μ0:3.5
 x̄:4.11
 Sx:2.485870415…
 n:30
 μ:≠μ0 <μ0 >μ0
 Calculate Draw
```

OUTPUT
```
T-Test
 μ≠3.5
 t=1.344039327
 p=.1893568545
 x̄=4.11
 Sx=2.485870416
 n=30
```

CONFIDENCE INTERVAL

INPUT
```
TInterval
 Inpt:Data Stats
 x̄:4.11
 Sx:2.485870415…
 n:30
 C-Level:.95
 Calculate
```

OUTPUT
```
TInterval
 (3.1818,5.0382)
 x̄=4.11
 Sx=2.485870416
 n=30
```

This means that we can assert with a 95% degree of confidence that the interval (3.1818… , 5.0382…) contains the true mean.

TWO SAMPLES

σ known

HYPOTHESIS $\mu_1 = \mu_2$

2-SampZTest

INPUT
```
2-SampZTest
 Inpt:Data Stats
 σ1:90
 σ2:90
 x̄1:647
 n1:391000
 x̄2:644
↓n2:9000
```

OUTPUT
```
2-SampZTest
 μ1≠μ2
 z=3.12649964
 p=.0017691417
 x̄1=647
 x̄2=644
↓n1=391000
```

CONFIDENCE INTERVAL – 2-SampZInt

INPUT
```
2-SampZInt
 Inpt:Data Stats
 σ1:90
 σ2:90
 x̄1:647
 n1:391000
 x̄2:644
 n2:9000
 C-Level:.95
 Calculate
```

OUTPUT
```
2-SampZInt
 (1.1193,4.8807)
 x̄1=647
 x̄2=644
 n1=391000
 n2=9000
```

This means that we can assert with a 95% degree of confidence that the interval (1.1193, 4.8807) contains the true value of $\mu_1 - \mu_2$. Note: 0 is not in this interval.

σ unknown

HYPOTHESIS $\mu_1 = \mu_2$

2-SampTTest

INPUT
```
2-SampTTest
 Inpt:Data Stats
 x̄1:647
 Sx1:1.28723172…
 n1:391000
 x̄2:644
 Sx2:.757005595…
↓n2:9000
```

OUTPUT
```
2-SampTTest
 μ1≠μ2
 t=220.2234862
 p=0
 df=399998
 x̄1=647
↓x̄2=644
```

CONFIDENCE INTERVAL – 2-SampTInt

INPUT
```
2-SampTInt
 Inpt:Data Stats
 x̄1:647
 Sx1:1.28723172…
 n1:391000
 x̄2:644
 Sx2:.757005595…
 n2:9000
 C-Level:.95
 Pooled:No Yes
 Calculate
```

OUTPUT
```
2-SampTInt
 (2.9733,3.0267)
 df=399998
 x̄1=647
 x̄2=644
 Sx1=1.28723173
↓Sx2=.757005595
```

This means that we can assert with a 95% degree of confidence that the interval (2.9733, 3.0267) contains the true value of $\mu_1 - \mu_2$. Note: 0 is not in this interval.

Copyright © 1998 by Brendan Kelly Publishing Inc.

EXPLORATION 17 Hypothesis Testing - 1 Sample

©1997 Reprinted with Special Permission of King Features Syndicate

The student in the cartoon has used past data to formulate a hypothesis about his chances of passing the statistics course. The ultimate test of his hypothesis will come when he receives his grade. In this unit you will discover how to test various hypotheses by using the properties of the distributions you studied in the previous unit.

WORKED EXAMPLE 1

John Kerrich, a mathematician visiting Denmark during World War II, was captured by the Germans and placed in a prison camp. During his internment he ran several probability experiments. In one experiment he tossed a coin 10,000 times. On the first 100 tosses, it turned up heads 44 times. After 10,000 tosses, it turned up heads 5,067 times.
Test the hypothesis: *The coin is fair; i.e. heads and tails are equally probable outcomes.*

a) After 100 tosses. b) After 10,000 tosses.

SOLUTION

a) The coin-tossing experiment satisfies the conditions of *Bernoulli trials* presented at the bottom of page 47. Therefore if H is the random variable representing the number of heads, then H has a binomial distribution. If H is truly a fair coin then any deviation from 50 heads on 100 tosses would have to be a result of *chance* variation. But Kerrich observed that there were only 44 heads – 6 heads short of the expected number. What is the probability that a fair coin would yield 6 or more heads above or below the number 50? That is, what is the probability that $H \le 44$ or $H \ge 56$?

If $F_H(x)$ denotes the cumulative probability function for H, then the probability that $H \le 44$ is $F_H(44)$ and the probability that $H \ge 56$ is $1 - F_H(55)$. Therefore the probability that $H \le 44$ or $H \ge 56$ is $1 - F_H(55) + F_H(44)$. or more conveniently $1 - [F_H(55) - F_H(44)]$. To compute the value of $[F_H(55) - F_H(44)]$, we could compute the sum of the values of the binomial probabilities ${}_{100}C_k\, 2^{-100}$ for $k = 45, \ldots, 55$. Alternatively, we use the binomial cumulative distribution function **binomcdf(** command on your TI-83. This is accessed by pressing: **2nd** [DISTR] **ALPHA** [A]. We then enter the commands as shown on the display to obtain 0.728… That is, the probability that H differs from 50 by more than 5 is 1 – 0.728 or about 0.27. This means that if the coin is fair, the heads would differ from the expected value by this much or more in only 27% of the cases. It's not enough to reject the hypothesis that the coin is fair, but it's not confirming evidence either!

probability of a head ↓

```
binomcdf(100,.5,
55)-binomcdf(100
,.5,44)
          .7287469781
```


Copyright © 1998 by Brendan Kelly Publishing Inc.

b) If we attempt the same approach as in part a), we obtain an error message (see Exercise 1), because the TI-83 cannot handle the computation of such large binomial coefficients. Consequently it uses the fact, discovered in *Exploration* 12, that for a large number of trials the binomial distribution approaches a normal distribution. The random variable H is approximated by a normally distributed random variable X with mean 5000 and standard deviation $\sqrt{2500}$ or 50. Therefore, we can test the hypothesis by evaluating the area under the graph of X and satisfying $X \le 4933$ or $X \ge 5067$.

> **Note**
> The binomial distribution associated with *n* trials for which the probability of a "success" is *p* has a mean of *np* and standard deviation $\sqrt{np(1-p)}$.

This area is given by $1 - [F_X(5067) - F_X(4933)]$, where F_X is the cumulative probability density function of X. To evaluate $[F_X(5067) - F_X(4933)]$, the shaded area under the graph and in the interval $4933 \le X \le 5067$, we use the **ShadeNorm(** command as shown in the display. This yields the area 0.819... and so the unshaded area is $1 - 0.819... \approx 0.180$. This indicates that *if the coin is fair*, the number of heads will only differ from the expected number by as much or more than 67 on about 18% of such experiments. Although this does not suggest that the hypothesis is true, it is *not* a small enough percentage to enable us to discard the hypothesis as highly improbable.

```
ShadeNorm(4933,5
067,5000,50)
```

$4800 \le x \le 5200$
$0 \le y \le 0.01$

Note: Part b) could have been solved using the statistical test **1-PropZTest** (one proportion Z-test) on your TI-83. This test is used to test the following "null" hypothesis about the value of a particular proportion *P*.

Any difference between the observed value of P and its expected value p_0, *is purely a chance variation.*

against any of the alternative hypotheses $P \ne p_0$, $P < p_0$, or $P > p_0$.

To access this test press: STAT ▶ ▶ 5 and enter the values p_0 (the expected proportion), x (the observed number of heads), 10000 (the number of trials), and $\ne p_0$ the alternative hypothesis, to obtain the display below left. You will obtain one of the displays below depending upon whether you select **Calculate** or **Draw**.

```
1-PropZTest
 p0:.5
 x:5067
 n:10000
 prop≠p0 <p0 >p0
 Calculate Draw
```

```
1-PropZTest
 prop≠.5
 z=1.34
 p=.1802454677
 p̂=.5067
 n=10000
```

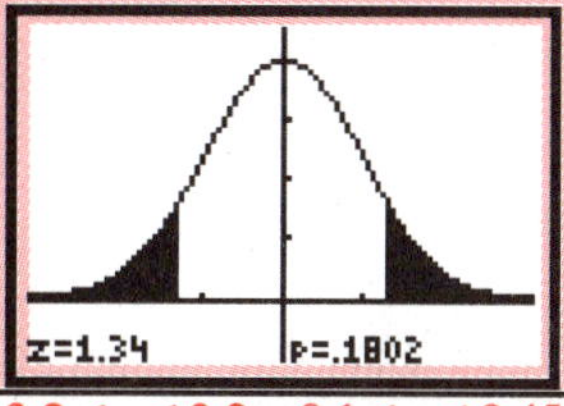

$-3.2 \le x \le 3.2$; $-0.1 \le y \le 0.45$

The middle display yields p = 0.180... as in part b).

The statistical tests that the TI-83 provides on its STAT menu are very useful for testing hypotheses; however, it is easy to enter the incorrect numbers into these menus if you do not understand the mathematics underlying each test. For this reason it is recommended that you tackle tests of significance without using these convenient menus until you have mastered the "distribution approach". When you are confident that you understand this approach, you should practice using the TI-83 statistical tests to verify your solutions.

Copyright © 1998 by Brendan Kelly Publishing Inc.

In *Worked Example* 1, the tossing of a coin 10,000 times was treated as if 10,000 identical coins were tossed once, and the number landing heads was counted. For this reason the binomial distribution of the number of heads could be approximated by a normal distribution with mean 5000 and standard deviation 50. Since most distributions cannot be approximated by a normal distribution, the method employed in *Worked Example* 1 cannot be used in general. However, the following remarkable theorem tells us that if we take a sample of size n from almost any population with mean μ and standard deviation σ, and take the mean of these values and then repeat this process many times, each time selecting samples of size n, then the values taken by the mean have a near-normal distribution with mean μ and standard deviation $\sigma/\sqrt{n}$. This is known as the *Central Limit Theorem.*

Theorem (Central Limit Theorem)

If $X_1, X_2, \ldots X_n$ are independent random variables whose values are taken from a non-normal population with mean μ and standard deviation σ, then the distribution of $\overline{X} = (X_1 + X_2 + \ldots + X_n)/n$ is approximately normal with mean μ and standard deviation $\sigma/\sqrt{n}$.

WORKED EXAMPLE 2

The Chief Operating Officer of a chain of supermarkets advises its store managers that the time spent by customers at the checkout counter (excluding time waiting in line) has a mean of 2.4 minutes and standard deviation 2 minutes. To test this assertion one of the managers observed that it took 3 hours and 52 minutes to process 80 customers (excluding time waiting in line).

Test the assertion of the Chief Operating Officer based on this single sample.

SOLUTION

Since the customers were not waiting in line, we can assume that all the customer times were independent random variables $X_1, X_2, \ldots X_{80}$. *If the Chief Operating Officer is correct*, the random variables $X_1, X_2, \ldots X_{80}$ are drawn from a population with mean 2.4 and standard deviation 2. Then the Central Limit Theorem implies that the sample mean $\overline{X}$ for samples of size 80 is a normally distributed (or near-normal) random variable with mean 2.4 and standard deviation $2/\sqrt{80} \approx 0.2236$. The mean checkout time per customer for the sample of 80 customers is $(3 \times 60 + 52) \div 80$ or 2.9 minutes.

We pose the null hypothesis: *The difference between the observed value of* $\overline{X}$ (2.9) *and the expected value* (2.4) *is entirely chance variation.*

To test this hypothesis, we must calculate the probability that a normally distributed variable with mean 2.4 and standard deviation$2/\sqrt{80}$, or 0.224, would deviate from 2.4 by 0.5 or more. That is, we must evaluate $1 - [\,F_{\overline{X}}(2.9) - F_{\overline{X}}(1.9)\,]$ where $F_{\overline{X}}(x)$ is the cumulative distribution function of $\overline{X}$. As in *Worked Example* 1, we use the **ShadeNorm(** command. We see $[\,F_{\overline{X}}(2.9) - F_{\overline{X}}(1.9)\,] \approx 0.974657$ so $1 - [\,F_{\overline{X}}(2.9) - F_{\overline{X}}(1.9)\,] \approx 0.02534$. That is, if the null hypothesis is true, then chance variation alone could account for a discrepancy as large or larger in only about 2.5% of the cases. We can *reject* the hypothesis at the commonly used 0.05 (5%) level of significance.

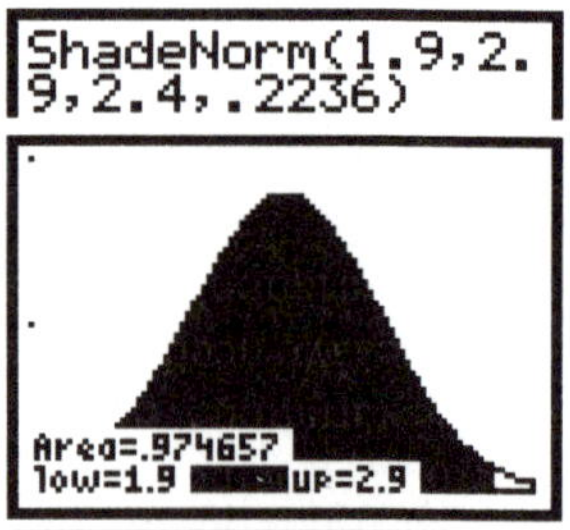

$1.8 \le x \le 3.0;\ 0 \le y \le 2$

As in *Worked Example* 1, we can use a statistical test as an alternative.

To access the Z-test, we press: STAT ▶ ▶ 1

We complete the menu as shown in the display below left and we obtain the other two displays according to whether we select **Calculate** or **Draw**.

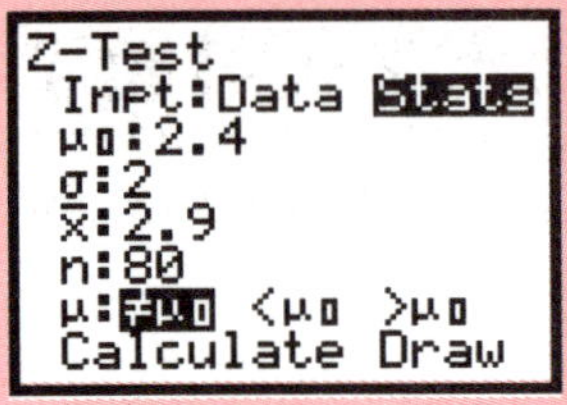

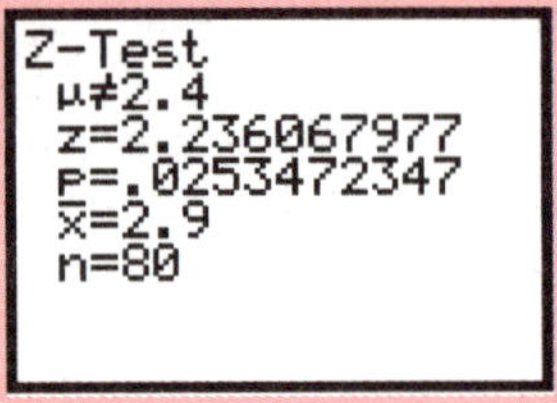

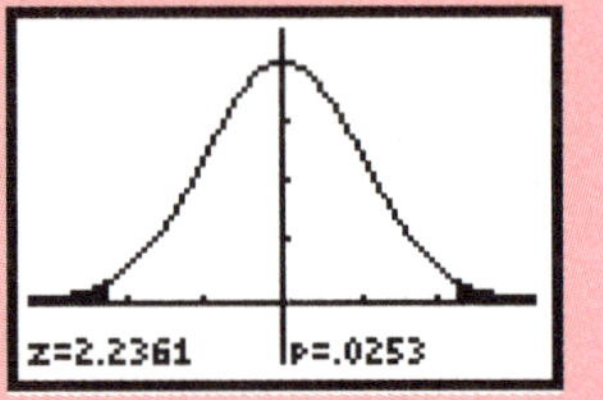

Copyright © 1998 by Brendan Kelly Publishing Inc.

Exercises

1. a) Use the method of *Worked Example* 1 a) to calculate the probability that 100 tosses of a fair coin will yield between 40 and 60 heads inclusive, i.e. the number of heads H satisfies the inequality $40 \le H \le 60$.

b) Use the method in part a) to calculate the probability that 10,000 tosses of a fair coin will yield H heads where $4933 \le H \le 5067$. Describe what you find. Why do you think this happens?

2. What is the probability that the number of heads on 500 tosses of a fair coin will yield more than 275 heads?

3. Solve Problem 2 by approximating the binomial distribution with a normal distribution. (See the marginal note at the top of page 69.)

4. In the solution to *Worked Example* 2, we assumed that $\overline{X}$ is normally distributed, even though we were not told that X_1, X_2, ... X_N are normally distributed. Was our assumption justified? Explain why or why not.

5. A fair coin is tossed 100 times. Find the smallest value of h such that there is a probability of 0.90 that the number of heads H satisfies the condition $50 - h \le H \le 50 + h$. (This is called the 90% *confidence interval*.) See the marginal note at the top of page 69.

6. a) A manufacturer advertises that its bulbs have a mean lifetime of 8000 hours. Previous measurements have ascertained that the standard deviation of these lifetimes is 150 hours. During a quality control inspection, 60 bulbs are tested and it is found that the mean lifetime of these bulbs is only 7952 hours. State the null hypothesis associated with this test against the hypothesis that the mean lifetime is not 8000 hours. Then test this hypothesis at the 0.05 and 0.01 levels of significance.

b) Test the null hypothesis against the hypothesis that the true mean lifetime is *less than* 8000.

7. A normal die has one spot on one face, two spots on another face, and so on up to the sixth face, which has six spots.

a) If the die is a fair die (all faces equally probable) what is the mean number of spots per toss that should show on the die?

b) What is the standard deviation of the number of spots that will show on the toss of the die?

c) The total number of spots that appear on 90 tosses of a particular die is 372. Test the hypothesis that this is a fair die (all faces equally probable) at the 0.01 level of significance.

Investigations

HOW DO YOU TEST A HYPOTHESIS WHEN THE STANDARD DEVIATION OF THE POPULATION IS UNKNOWN?

8. In the previous exercises, you tested hypotheses about the mean of a population for which the standard deviation was known. In this investigation, you will discover how statisticians test hypotheses about the mean of a population when its standard deviation is unknown.

When the standard deviation of a population is unknown, we must estimate its value using a sample of the population. The estimator used for the standard deviation of a population is the *standard deviation* of the sample defined by:

$$s = \sqrt{\frac{\sum_{i=1}^{n}(x_i - \overline{x})^2}{n-1}}$$

Observe that the standard deviation of a sample resembles the standard deviation of a population except the denominator is $n - 1$ instead of n.

a) If X is a normally distributed random variable with mean μ and standard deviation σ, then $(X - \mu)/\sigma$ is a random variable with mean 0 and standard deviation 1. Use this fact and the statement at the bottom of page 65 to prove that if $\overline{X}$ denotes the mean of a sample of size n taken from a normal population with mean μ and standard deviation σ, then the random variable $\sqrt{n}(\overline{X} - \mu)/\sigma$ is normally distributed with mean 0 and standard deviation 1.

b) Use the theorem at the bottom of page 64 to prove that $(n-1)s^2/\sigma^2$, where s is the standard deviation defined above, has a chi-square distribution with $n - 1$ degrees of freedom.

c) Use parts a) and b) above and the theorem under the heading "For Calculus Buffs" on page 66 to show that the random variable T given by:

$$T = \frac{\overline{X} - \mu}{s/\sqrt{n}}$$

has a t distribution with $n - 1$ degrees of freedom.

d) The manufacturer of Brand A batteries (see page 19) asserts that the mean life of its batteries is 6.2 years. Use the sample of the 30 Brand A battery lifetimes given on page 19 and the T statistic given in part c) to test the manufacturer's claim at a 5% level of significance.

e) Verify your answer in part d) by using the T-Test template accessed by pressing STAT ▶ ▶ 2

f) Repeat the hypothesis test at the 95% confidence level using the Z-test instead of the T-test. Explain what you discover.

Copyright © 1998 by Brendan Kelly Publishing Inc.

A-2 THE GAZETTE

State SAT Scores in Math Below National Average

BY ANDREA WELBURN
Education Editor

Newly released SAT scores indicate that state-wide performance of 9,000 of our students in the 1997-98 Math IIC Test fell 3 points below the national average. This test administered to approximately 400,000 college-bound senior students nation-wide, is regarded as an indicator of the level of preparation of students at the end of high school. The national mean score on this test was 647. The standard deviation, a measure which statisticians use to indicate the spread in data, was 90 nationally. The city's Superintendent J. Ross explained that while the scores were somewhat disappointing, a difference of only three points on a mean score of 647 is not statistically significant.

This article raises the important question – *When is a difference between the means of two populations statistically significant?* The students in this state had a mean score of 644 (i.e. 3 points below the national average of 647).

To set up a test of significance, we consider a slightly different but related problem. Suppose X_1 and X_2 are both normally distributed random variables, with means μ_1 and μ_2 and standard deviations σ_1 and σ_2 respectively. Then $\overline{X}_1$ the mean of a sample of n_1 values of X_1 and $\overline{X}_2$ the mean of a sample of n_2 values of X_2 are normally distributed with means μ_1 and μ_2 and standard deviations $\sigma_1/\sqrt{n_1}$ and $\sigma_2/\sqrt{n_2}$.
What test statistic will enable us to test the hypothesis $\mu_1 = \mu_2$?

The distribution of the statistic $\overline{X}_1 - \overline{X}_2$ is normal with mean $\mu_1 - \mu_2$ and standard deviation $\sigma_{\overline{X}_1 - \overline{X}_2}$ where

$$\sigma_{\overline{X}_1 - \overline{X}_2} = \sqrt{\sigma_{\overline{X}_1}^{\,2} + \sigma_{\overline{X}_2}^{\,2}} = \sqrt{\frac{\sigma_1^2}{n_1} + \frac{\sigma_2^2}{n_2}}$$

Therefore the statistic $Z = \dfrac{\overline{X}_1 - \overline{X}_2}{\sqrt{\dfrac{\sigma_1^2}{n_1} + \dfrac{\sigma_2^2}{n_2}}}$ is normally distributed with mean $\mu_1 - \mu_2$ and standard deviation 1. The null hypothesis, $\mu_1 = \mu_2$ or equivalently $\mu_1 - \mu_2 = 0$, can therefore be tested by testing the hypothesis that the mean value of Z is zero. Using Z as our test statistic reduces the testing of the hypothesis that two populations have equal means to a test of the hypothesis that the mean of Z is zero.

WORKED EXAMPLE 1

Use the information given in the article above to determine whether the difference in the means of the state-wide and national SAT scores is significant.

SOLUTION

If the SAT scores are an indicator of the level of mathematical achievement of American seniors, then the national and state scores on these tests might be regarded as samples of size $n_1 = 391{,}000$ (out-of-state entries) and size $n_2 = 9{,}000$ (in-state entries) from the entire population of college-bound seniors. These samples can be regarded as drawings from normally distributed variables X_1 and X_2 with means μ_1 and μ_2 and standard deviations σ_1 and σ_2 respectively. We want to test the hypothesis $\mu_1 - \mu_2 = 0$.

The test of significance developed above applies to two *independent* samples. The samples in the article are dependent since the state scores formed part of the national data. Furthermore, we are given the standard deviation for the national but not the state sample. Therefore to apply the foregoing test of significance to this problem, we will have to make two simplifying assumptions.

1. Since n_1 and n_2 are large we assume that $\sigma_1 = \sigma_2 = 90$.
2. Since n_1 is much larger the n_2, we will not adjust the mean score $\overline{X}_1$ of the national population to account for the in-state scores that were included.

Substituting $\overline{X}_1 - \overline{X}_2 = 3$ and $\sigma_1 = \sigma_2 = 90$ and $n_1 = 391{,}000$ and $n_2 = 9000$ into the expression for Z, we find $Z = 3.1265$. But Prob $(|Z| \geq 3.1265) = 1 -$ **normalcdf(-3.1265,3.1265)** $= 0.00176\ldots$ i.e., the probability that the difference in the means is a chance variation is less than 1%, so the difference is significant at the 0.01 level of significance.

```
1-normalcdf(-3.1
265,3.1265)
      .0017691395
```


Copyright © 1998 by Brendan Kelly Publishing Inc.

Worked Example 1 could have been solved using the 2-Sample Z-test template..
To access the 2-Sample Z-test, we press: STAT ▶ ▶ 3
We complete the menu as shown in the display below left and we obtain the other two displays according to whether we select **Calculate** or **Draw**.

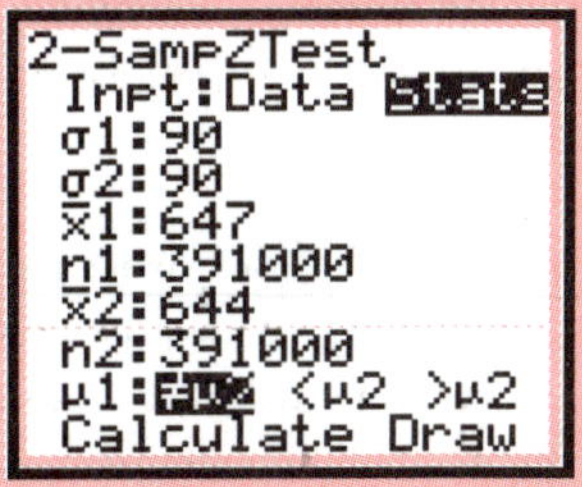

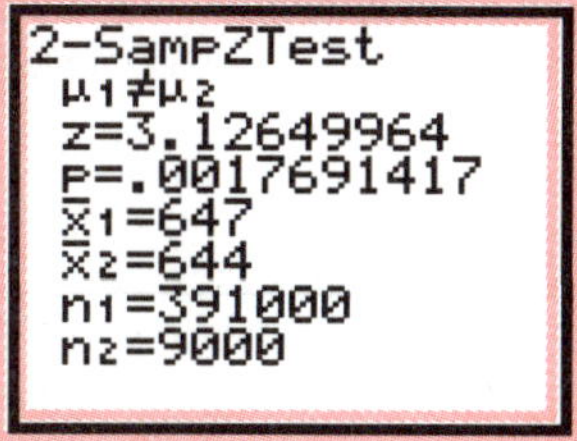

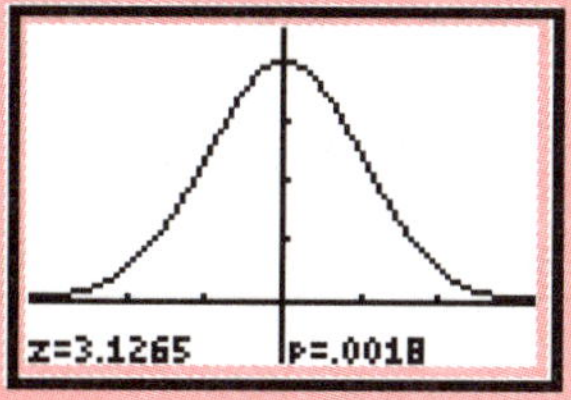

The displays show the input statistics and the output, verifying the answer given.

The 2-Sample Z-test is applicable when the standard deviations of the two samples are known. However in most situations the standard deviations σ_1 and σ_2 are unknown, and so we must use the standard deviations s_1 and s_2 of the samples to estimate them. If we assume that the two populations have the same standard deviation (called the *homoscedacity assumption*), i.e. $\sigma_1 = \sigma_2 = \sigma$, then we can *combine* both s_1 and s_2 into a "pooled" estimate s_p of the standard deviation σ, where s_p is given below. To test the hypothesis that the means of two populations are equal when their standard deviations are unknown, we replace σ in the 2-Sample Z-test with the test statistic

$$T = \frac{\overline{X}_1 - \overline{X}_2}{s_P\sqrt{\frac{1}{n_1}+\frac{1}{n_2}}} \quad \text{where } s_P = \sqrt{\frac{(n_1-1)s_1^{\,2}+(n_2-1)s_2^{\,2}}{n_1+n_2-2}}$$

The pooled estimate s_p is the standard deviation we obtain by adding the sum of the squares of the "within-sample" deviations from the mean for both samples, dividing by the number of degrees of freedom $n_1 + n_2 - 2$ and then taking the square root. In Investigation 9, page 74, you will show that the statistic T has a t distribution with $n_1 + n_2 - 2$ degrees of freedom. *Worked Example* 2 shows how we can apply this statistic to test the hypothesis that two populations have equal means.

A recent research study by *Wood and Sellers*[1] sought to determine whether third-grade students exposed to a problem-solving approach to mathematics instruction over two years would show any difference in their understanding of mathematical concepts from those in traditional instructional environments. The researchers divided the classes into two groups. The *treatment group* received the problem-centered mathematics instruction while the *control group* received traditional instruction. At the end of the experimental period, the mean scores of the two groups on a concept test were compared to determine whether there was a significant difference in performance. The following example shows their data and the test of significance they used.

WORKED EXAMPLE 2

The mean scores with standard deviations attained on the concepts test by the 6 classes in each group are shown in the table. Test, at the 1% level of significance, the hypothesis that the difference in the means is purely a chance variation.

Group	Mean	Standard Deviation
Treatment	8.24	0.32
Control	6.58	0.33

SOLUTION

We substitute the following values into the test statistic T given above:

$n_1 = n_2 = 6;\ \bar{x}_1 = 8.24,\ s_1 = 0.32;\ \bar{x}_2 = 6.58,\ s_2 = 0.33.$

This yields $T \approx 8.85$. To test the null hypothesis $\mu_1 - \mu_2 = 0$, we evaluate the area contained in the two tails of the t distribution with 10 degrees of freedom.

Prob ($|T| \geq 8.85$) = 1 – **tcdf(-8.85, 8.85,10)** $\approx 4.8 \times 10^{-6}$.

This shows why the researchers rejected the null hypothesis at the 1% level of significance.

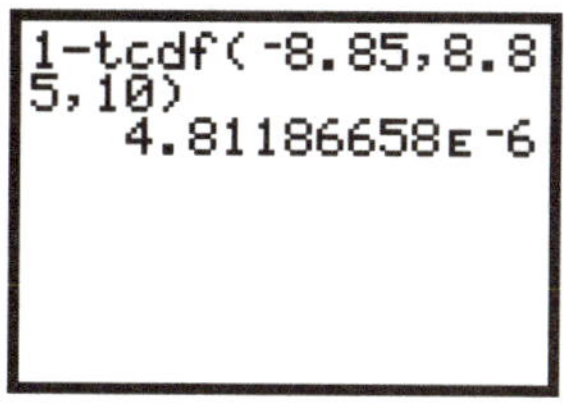

[1] Wood, T., & Sellers, P. (1996). Assessment of a Problem-Centered Mathematics Program. *Journal for Research in Mathematics Education, 27, 337-353.*

Copyright © 1998 by Brendan Kelly Publishing Inc.

Exercises

1. Is it more likely that the null hypothesis will be rejected when testing at the 5% level of significance or when testing at the 1% level of significance? Explain your answer.

2. In the newspaper article on page 72, Superintendent Ross stated that a 3-point difference on a mean score of 647 is not statistically significant. Explain the difference in meaning of the terms "statistically significant" and "significant".
a) Do you think Ross's statement is correct? Why?
b) Would you agree with Ross's statement if he had said "significant" rather than "statistically significant"? Explain.

3. Explain how you decide whether to use a 2-Sample Z-test or a 2-Sample T-test when testing whether the means of two populations are the same.

4. a) In *Worked Example* 2, calculate the pooled standard deviation s_p for the treatment and control groups. How could you have predicted this value without computation?
b) The 2-Sample *t*-Test was used to test the null hypothesis because the standard deviation of the population was unknown. Apply the 2-Sample Z-test to the null hypothesis using the value of s_p you calculated in part a) as the "known" value of σ. Compare this result with the 2-Sample T-test.

5. The lifetimes (in years) of a sample of 30 Brand A and 30 Brand B batteries are given in the tables below. Test the hypothesis that the two brands have the same mean lifetime, at the 5% level of significance.

Measured Lifetimes of 30 Brand A Batteries				
5.7	7.3	6.9	5.7	5.6
6.2	6.7	5.5	4.9	6.9
6.0	4.8	4.1	5.3	8.3
6.3	7.5	5.0	5.7	9.3
3.9	3.1	4.8	5.9	6.6
6.4	5.0	6.1	4.6	5.7

Measured Lifetimes of 30 Brand B Batteries				
5.4	6.3	4.0	5.9	5.6
4.7	5.1	3.3	6.0	5.3
5.0	4.5	5.8	4.4	4.9
5.7	6.2	5.6	4.9	6.0
4.9	5.7	6.2	4.5	5.8
6.8	5.9	5.3	5.6	5.9

6. To compare the abilities of students in different schools, the I.Q.'s of a random sample of students from Lincoln High and Bloomfield Secondary School were measured. The table summarizes the findings. Test the hypothesis that there is a significant difference in the I.Q.'s at those two schools.
a) at the 1% b) at the 5% level of significance.

School	Number of Students	Mean IQ	Standard Deviation
Lincoln	18	116	8
Bloomfield	13	108	12

7. To determine whether graphing calculators help students solve math problems, a teacher divided her class into two randomly-chosen groups of 14. She gave 10 problems to both groups, allowing only one group to use the calculators. The mean scores of the calculator and control groups were respectively 76 and 71 with corresponding standard deviations of 7 and 9. Was there a significant difference in the performance of the groups?

Investigations

HOW DO YOU TEST A HYPOTHESIS THAT ASSERTS THAT PROPORTIONS IN TWO DIFFERENT SAMPLES ARE EQUAL?

8. A survey is conducted to determine whether males are more likely than females to access the Internet. The table summarizes the results.

	sample size	number who use internet
males	220	172
females	360	305

a) Suppose the true proportion p of all males who use the Internet is the same as the proportion of females who use it. If X_1 and X_2 denote the number of males and females in samples of size n_1 and n_2 respectively who report using the Internet, what is the distribution, mean, and standard deviation of the statistic $\dfrac{X_1}{n_1} - \dfrac{X_2}{n_2}$?

b) Use part a) to obtain a statistic that is normally distributed and has mean 0 and standard deviation 1.

c) Use your statistic in c) to test the hypothesis that there is no significant difference by sex in the use of Internet. Use the **2-PropZtest** on your TI-83 to verify your answer.

CHALLENGE

9. In this challenging investigation, you will combine several theorems to prove that T defined on page 73 has a t-distribution with $n_1 + n_2 - 2$ degrees of freedom.

a) Show that if $\overline{X}_1$ and $\overline{X}_2$ are sample means of random samples from normally distributed populations with means $\mu_1 = \mu_2 = \mu$ and standard deviation σ, then the random variable

$$\frac{\overline{X}_1 - \overline{X}_2}{\sigma\sqrt{\dfrac{1}{n_1} + \dfrac{1}{n_2}}}$$

is normally distributed with mean 0 and standard deviation 1.

b) Show the if s_p is defined as on page 73, then $\dfrac{(n_1 + n_2 - 2)s_p^2}{\sigma^2}$ has a chi-square distribution with $n_1 + n_2 - 2$ degrees of freedom.

c) Use parts a) and b) and the theorem on page 66 under ***FOR CALCULUS BUFFS*** to prove that if

$$T = \frac{\overline{X}_1 - \overline{X}_2}{s_P\sqrt{\dfrac{1}{n_1} + \dfrac{1}{n_2}}} \quad \text{where } s_P = \sqrt{\frac{(n_1 - 1)s_1^2 + (n_2 - 1)s_2^2}{n_1 + n_2 - 2}}$$

then T has a t distribution of $n_1 + n_2 - 2$ degrees of freedom

Copyright © 1998 by Brendan Kelly Publishing Inc.

EXPLORATION 19 The Chi-Square Test

cathy®

CATHY © Cathy Guisewite. Reprinted with permission of UNIVERSAL PRESS SYNDICATE. All rights reserved.

Cathy has made an inferential leap that provides us with food for thought. Weighty problems like Cathy's can sometimes be reduced to computation using a "goodness of fit" test such as the so-called *Chi-Square Test* presented below. (It's derivation is developed in the ***CHALLENGE*** exercise on page 77.)

In *Exploration* 17, we used the normal distribution to determine whether a coin is fair. However, this method worked because there are only two possible outcomes when a coin is tossed. How could we determine whether a die is fair? If a die is fair, all outcomes are equally likely. However, we know that chance variation will result in a discrepancy between the observed and expected frequencies of each outcome. We need a statistical device for deciding when this discrepancy is too large to attribute to chance variation alone. The chi-square test is such a device.

> Suppose a probability experiment has a set of k possible outcomes, $O_1, O_2, \ldots, O_k$ with expected frequencies $e_1, e_2, \ldots, e_k$. If $f_1, f_2, \ldots, f_k$ denote the corresponding observed frequencies, and $\sum_{i=1}^{k} f_i = \sum_{i=1}^{k} e_i = n$, then the statistic χ^2 given by $\chi^2 = \sum_{i=1}^{k} \frac{(f_i - e_i)^2}{e_i}$ has a chi-square distribution with $k - 1$ degrees of freedom.

WORKED EXAMPLE 1

To determine whether a die is fair (i.e. all outcomes equally likely) a casino inspector tosses it 120 times and obtains these frequencies for the possible outcomes.

Face turned up	⚀	⚁	⚂	⚃	⚄	⚅
Frequency	**26**	**15**	**24**	**23**	**17**	**15**

Test the hypothesis that the die is fair at the 5% level of significance.

SOLUTION

```
χ²cdf(6,E99,5)
      .3062189178
```

If f_i is the frequency of the outcome showing i spots, then the observed frequencies $f_1, f_2, \ldots, f_6$ are respectively 26, 15, 24, 23, 17, 15. Furthermore, the expected frequencies are $e_1 = e_2 = \ldots = e_6 = 20$. The chi-square test asserts that

$$\frac{(26-20)^2}{20} + \frac{(15-20)^2}{20} + \frac{(24-20)^2}{20} + \frac{(23-20)^2}{20} + \frac{(17-20)^2}{20} + \frac{(15-20)^2}{20} = 6$$

is the value taken by a statistic with a chi-square distribution with 5 degrees of freedom. To test whether this value is too large, we compute the area in the right tail of the graph of $\chi^2_5(x)$ as shown in the display. The display shows that about 30% of the time, chance variation will lead to comparable results or results less consistent with the hypothesis that the die is fair. Clearly we cannot reject this hypothesis at the 5% level of significance.

The chi-square test in *Worked Example* 1 measures the extent to which the observed frequencies f_i are proportional to the expected frequencies e_i under the constraint $f_1 + f_2 + \ldots + f_6 = e_1 + e_2 + \ldots + e_6 = n$ (the number of observations). We can generalize this model to test the hypothesis that two loaded dice are equivalent, that is, the probability of the outcome ⚀ for Die 1 is the same as the probability of that outcome for Die 2 and so on for each outcome.

observed frequencies	f_1	f_2	f_3	f_4	f_5	f_6
expected frequencies	e_1	e_2	e_3	e_4	e_5	e_6

Suppose we roll both dice n times and let $f_{11}, f_{12}, \ldots, f_{16}$ denote the observed frequencies of each outcome for Die 1, and $f_{21}, f_{22}, \ldots, f_{26}$ the corresponding frequencies for Die 2. We can display these frequencies in what is called a 2×6 *contingency table.* Since the dice are loaded, we do not know what the expected frequencies are, although our null hypothesis asserts they are the same for both dice. To estimate the expected frequencies e_i, we use our observations to weight the number in each cell according to the row and column to which it belongs. If R_i denotes the sum of the frequencies in the ith row, then we apply the weighting factor $R_i / \sum_{i=1}^{2} R_i$ to all the cells in that row. Similarly, if C_j denotes the sum of the frequencies in the jth column, we apply the weighting factor $C_j / \sum_{i=1}^{6} C_i$ to all the cells in that column. Therefore, the weighting factor that applies to the cell in the ith row and j^{th} column is $R_i / \sum_{i=1}^{2} R_i \times C_j / \sum_{i=1}^{6} C_i$

f_{11}	f_{12}	f_{13}	f_{14}	f_{15}	f_{16}
f_{21}	f_{22}	f_{23}	f_{24}	f_{25}	f_{26}

$$\sum_{j=1}^{6} f_{1i} = \sum_{j=1}^{6} f_{2i} = n$$

Then e_{ij}, the expected frequency in that cell, is the weighting factor times the sum of the frequencies in all the cells. That is,

$$e_{ij} = \frac{R_i C_j}{\sum_{i=1}^{2} R_i \sum_{i=1}^{6} C_i} \sum_{i=1}^{2} R_i = \frac{R_i C_j}{\sum_{i=1}^{6} C_i}$$

Once we have calculated all the e_{ij}, we test the hypothesis using the statistic $\chi^2 = \sum_{i=1}^{2} \sum_{j=1}^{6} \frac{(f_{ij} - e_{ij})^2}{e_{ij}}$.

This statistic has a chi-square distribution with $(r - 1)(k - 1)$ degrees of freedom, where r and k are respectively the number of rows and columns in the table. The following example shows how we can test variables for independence.

WORKED EXAMPLE 2

A researcher tests the hypothesis that there was no significant difference in the amount of graphing calculator use demanded by the different tests given to the three senior classes at Roosevelt High. To this end, she analyzed each of the three 50-item tests and classified each item as "inactive", "neutral", or "active" depending on the extent of calculator use required. Use the tallies shown in the 3×3 contingency table to test the hypothesis stated above.

	Test A	Test B	Test C
Inactive	16	19	13
Neutral	14	10	26
Active	20	21	11

SOLUTION

Testing for equivalence of the tests is the same as testing whether the proportion of test items in each category are approximately the same for each test. Therefore, we apply the χ^2 test described above with $r = c = 3$, i.e. with $(r - 1)(c - 1)$ or 4 degrees of freedom. To apply this test we first complete all the row totals R_i and the column totals C_j. We find $R_1 = 48$, $R_2 = 50$, and $R_3 = 52$. Also $C_1 = C_2 = C_3 = 50$. Then we calculate the expected frequencies e_{ij}, using the weighting factor $e_{ij} = \frac{R_i C_j}{150}$ to obtain the matrix of expected frequencies shown. We use these values to compute the matrix of terms $(f_{ij} - e_{ij})^2 / e_{ij}$ shown above. Adding all the elements of this matrix yields the value 12.945 for the test statistic $\chi_4^2 = \sum_{i=1}^{3} \sum_{j=1}^{3} \frac{(f_{ij} - e_{ij})^2}{e_{ij}}$.

Expected Frequencies e_{ij}		
16	16	16
50/3	50/3	50/3
52/3	52/3	52/3

Matrix of Terms $(f_{ij} - e_{ij})^2 / e_{ij}$		
0	9/16	9/16
32/75	8/3	$28^2/150$
16/39	$11^2/156$	$19^2/156$

As the display shows, the probability that the distribution χ^2_4 takes a value this large or larger is about 0.0115, so the researcher would reject the null hypothesis at the 5% level but not at the 1% level of significance. In Investigation 8, p. 77, you will discover how to use the χ^2 test on your TI-83 to perform all these computations once the contingency table is entered.

```
χ²cdf(12.945,E99
,4)
         .011547684
```


Copyright © 1998 by Brendan Kelly Publishing Inc.

Exercises

1. Explain why we cannot use the binomial distribution (as we did in *Worked Example* 1, p. 68) to determine whether a die is fair.

2. a) How do we find the expected frequencies when testing a die to determine if it is fair?

b) How do we find the expected frequencies to determine whether two loaded dice are "equivalent" (i.e. have the same set of probabilities for corresponding outcomes.)?

3. The observed frequencies for 210 tosses of a die are shown.

Face turned up	⚀	⚁	⚂	⚃	⚄	⚅
Frequency	**42**	**31**	**45**	**23**	**29**	**40**

Test, at the 1% level of significance, the hypothesis that the die is fair.

4. A survey of 256 families with exactly 4 children tallied the number of families having each combination of boys and girls shown in the table.

4 girls 0 boys	3 girls 1 boy	2 girls 2 boys	1 girl 3 boys	0 girls 4 boys
11	40	94	76	35

a) Calculate the expected frequency for each cell.

b) Test the hypothesis at the 1% level of significance that both sexes are equally likely.

5. The *Health and Nutrition Examination Survey* (HANES) of 1976–1980 sampled 2237 Americans and recorded for both sexes the number of people who were right-handed, left-handed, and ambidextrous. Use the data in the table to test the hypothesis that "handedness" is independent of sex. That is, there are the same proportion of males as females in each category.

	Women	Men
Right-handed	1070	934
Left-handed	92	113
Ambidextrous	8	20

6. A random sample of 2000 people were asked the question, "Do you think smoking causes cancer?". The respondents were asked whether or not they were smokers. Their answers are summarized in the table. Test at the 5% level the hypothesis that the answer to the question does not depend upon whether or not the respondent smokes.

	Yes	No	No Opinion
Non-Smokers	1321	32	17
Smokers	501	107	22

7. Solve *Worked Example* 1 *p. 68* using a χ^2 test and compare the results of both methods.

Investigations

HOW DO YOU PERFORM A CHI-SQUARE TEST USING THE TEST MENU ON YOUR TI-83?

8. In this investigation, you will discover how to use the test menu on your TI-83 to perform the chi-square test.

a) To solve *Worked Example* 2 using your TI-83, enter the numbers in the contingency table into the matrix [A] by pressing these keys: [MATRIX] [▶] [▶] [ENTER]

Enter 3×3 to define [A] as a 3×3 matrix. Then enter the elements of [A] as shown in the display.

```
MATRIX[A] 3 ×3
[ 16    19    13    ]
[ 14    10    26    ]
[ 20    21    11    ]
```

b) To access the χ^2 test, press:

[STAT] [▶] [▶] [ALPHA] [C]

You will see a display like this. Press [ENTER] twice to select [A] as the matrix of observed frequencies, [B] as the matrix in which the expected frequencies will be stored, and to choose the "calculate" option. Then press [ENTER] to obtain the final display of the χ^2 calculation. Does it agree with the results presented in *Worked Example* 2?

```
χ²-Test
 Observed:[A]
 Expected:[B]
 Calculate Draw
```

9. Use the χ^2 test on your TI-83 (see *Investigation* 8 above) to verify the solution to *Worked Example* 1, by following these steps.

a) We wish to compare the observed frequencies of the die we are testing with the expected frequencies of a fair die to see if the frequencies of the tested die are in similar proportion to the frequencies of a fair die. Therefore, we enter the observed frequencies 26, 15, 24, 23, 17, and 15 into the first row of a 2×6 matrix[A] and enter 1E12 into each entry of the second row as shown. (A fair die rolled 6 trillion times would have these expected frequencies.)

```
MATRIX[A] 2 ×6
[ 26    15    24    23    17    15   ]
[ 1E12  1E12  1E12  1E12  1E12  1E12 ]
```

b) Access the TI-83 χ^2 test as in 8b) and verify that it yields the same value for the χ^2 statistic as in *Worked Example* 1.

CHALLENGE

Let X_i denote the number of times that i spots appear on n tosses of a die. Show that for large n:

a) X_i is normally distributed with $\mu = np$ and $\sigma = \sqrt{np(1-p)}$ where p is the probability of that outcome on each toss.

b) The variable $\dfrac{X_i - np}{\sqrt{np(1-p)}}$ is normal with $\mu = 0$ and $\sigma = 1$.

c) $\chi^2 = \sum_{j=1}^{6} \dfrac{(X_i - np)^2}{np}$ has a χ^2 distribution.

Copyright © 1998 by Brendan Kelly Publishing Inc.

EXPLORATION 20 *Analysis of Variance*

In *Exploration* 19, Cathy made a giant inferential leap and concluded that doing math makes us fat. She might have verified her conclusion using the "goodness of fit" test shown on the right. As the quest to monitor body mass increases, the investigation of causes of weight gain intensifies. One experiment to determine which of the three diets, the low calorie diet (LOWCAL), the low carbohydrate diet (LOWCARB), or the low fat diet (LOWFAT), would result in the greatest weight loss was conducted by the *Statistically Significant Weight-Loss Clinic*[1].

The researcher who designed this scientific experiment chose 3 groups of 16 people at random from the weight-loss clinic. Each group was placed on one of the three diets for 6 months. The weight loss of each member was recorded and the mean weight loss for each group was compared. The data collected is shown in the table below.

It is not immediately clear how to determine whether the differences in the mean weight losses between groups is a chance variation. In *Exploration* 18, we saw how to use the *t*-distribution to test for a significant difference between two means, but this technique does not work when comparing differences among more than two means. To test the null hypothesis that there is no significant difference in the mean weight losses in the three treatment groups, i.e. $\mu_1 = \mu_2 = \mu_3$, the researcher used an important statistical technique called the *analysis of variance* (ANOVA).

6-Month Weight Loss (lbs.)		
LOWCAL	**LOWCARB**	**LOWFAT**
9	2	36
5	8	15
11	11	–2 (gain)
8	3	1
30	23	9
17	7	4
9	18	11
–8 (gain)	–6 (gain)	13
34	13	22
5	4	2
–3 (gain)	14	17
15	6	–8 (gain)
7	9	1
14	7	10
6	–5 (gain)	9
9	6	4
$\bar{x}_1 = 10.5$	$\bar{x}_2 = 7.5$	$\bar{x}_3 = 9$

The ANOVA procedure expresses the total variation in the data as a sum of the within-sample variation and the between-sample variation. We show below that if the null hypothesis is true, the ratio of the between-sample variation to the within-sample variation has an F-distribution. This enables us to test the null hypothesis by calculating that ratio. If the ratio is close to 1, it suggests that the differences in the means is attributable to chance variation, but if the ratio is substantially larger than 1, then the data suggest that the null hypothesis should be rejected. We express these ideas mathematically as follows.

Suppose we consider three treatment groups having *n* people per group. Let $L_1 = \{x_{11}, x_{12}, \ldots, x_{1n}\}$, $L_2 = \{x_{21}, x_{22}, \ldots, x_{2n}\}$, and $L_3 = \{x_{31}, x_{32}, \ldots, x_{3n}\}$ denote the weight losses of the *n* people in the three groups. The total squared deviations of all the weight losses in all groups combined is:

$$SS_T = \sum_{i=1}^{3}\sum_{j=1}^{n}(x_{ij} - \bar{x})^2 \quad \text{where } \bar{x} \text{ is the mean of all the data.}$$

The total squared deviations from the means within the samples is:

$$SS_w = \sum_{i=1}^{3}\sum_{j=1}^{n}(x_{ij} - \bar{x}_i)^2 \quad \text{where } \bar{x}_i \text{ is the mean for the } i^{th} \text{group.}$$

The total squared deviations of the means between samples is:

$$SS_b = \sum_{i=1}^{3}\sum_{j=1}^{n}(\bar{x}_i - \bar{x})^2 \quad \text{where } \bar{x}_i \text{ is the mean for the } i^{th} \text{group.}$$

Basic algebra (see *Investigation* 10 p. 80) yields $SS_T = SS_w + SS_b$. That is,

$$\sum_{i=1}^{3}\sum_{j=1}^{n}(x_{ij} - \bar{x})^2 = \sum_{i=1}^{3}\sum_{j=1}^{n}(x_{ij} - \bar{x}_i)^2 + \sum_{i=1}^{3}\sum_{j=1}^{n}(\bar{x}_i - \bar{x})^2$$

SS_T Total Sum of the Squares; SS_w Error Sum of the Squares; SS_b Between Samples Sum of the Squares

[1] This experiment is purely hypothetical. However, the author strongly recommends a steady diet of the ANOVA procedure.

Copyright © 1998 by Brendan Kelly Publishing Inc.

If the null hypothesis is true, then the x_{ij} can be regarded as a random sample from a normal population of mean μ and standard deviation σ. Therefore, the random variable

$$\frac{1}{\sigma^2}\sum_{i=1}^{3}\sum_{j=1}^{n}(x_{ij}-\bar{x})^2 = \frac{SS_T}{\sigma^2}$$

has a chi-square distribution with $3n - 1$ degrees of freedom (see Exercise 7). Similarly, we observe that

$$\frac{1}{\sigma^2}\sum_{i=1}^{3}\sum_{j=1}^{n}(x_{ij}-\bar{x}_i)^2 = \frac{SS_w}{\sigma^2} \qquad ①$$

has a chi-square distribution with $3(n - 1)$ degrees of freedom. Finally, we observe that

$\sum_{j=1}^{n}(\bar{x}_i-\bar{x})^2$ is independent of j, so $\sum_{j=1}^{n}(\bar{x}_i-\bar{x})^2 = n(\bar{x}_i-\bar{x})^2$. ②

Therefore, $\frac{1}{\sigma^2}\sum_{i=1}^{3}\sum_{j=1}^{n}(\bar{x}_i-\bar{x})^2 = \frac{n}{\sigma^2}\sum_{i=1}^{3}(\bar{x}_i-\bar{x})^2 = \frac{SS_b}{\sigma^2}$ has a chi-square distribution with 2 degrees of freedom. If we assume that random variables SS_w/σ^2 and SS_b/σ^2 are independent, it follows from the theorem on page 65, and equations ① and ② above, that the statistic

$\frac{SS_b/2\sigma^2}{SS_w/3(n-1)\sigma^2}$ or, simply $\frac{3(n-1)SS_b}{2SS_w}$ has an F distribution with 2 and $3(n - 1)$ degrees of freedom. Therefore, to test the null hypothesis $\mu_1 = \mu_2 = \mu_3$, we calculate $\frac{3(n-1)SS_b}{2SS_w}$ and apply the F test, as in the following example.

WORKED EXAMPLE

Use the analysis of variance method to test the null hypothesis $\mu_1 = \mu_2 = \mu_3$ at the 5% level of significance for the three weight-loss diets shown in the table on page 78.

SOLUTION

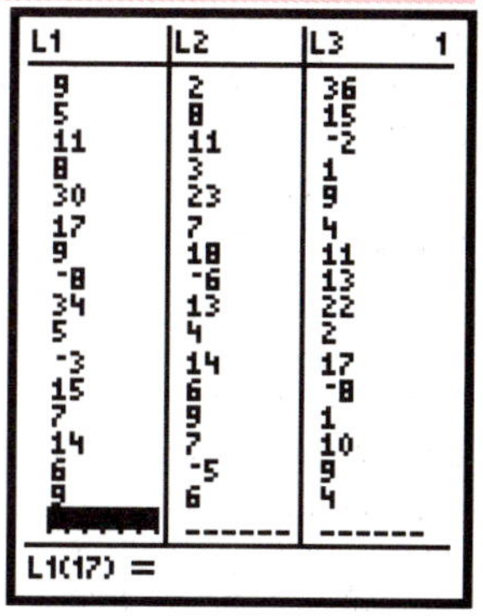

We enter the three columns of data LOWCAL, LOWCARB, and LOWFAT into the lists L_1, L_2 and L_3 respectively, by pressing STAT ENTER and entering each number as shown in the display. To evaluate SS_w and SS_b we must find $\bar{x}_i$ and s_i for $i = 1, 2,$ and 3. To obtain these values for $i = 1$, we press STAT ▶ ENTER 2nd [L_1] ENTER. We obtain the display below the list. We record $\bar{x}_1 = 10.5$ and $s_1 = 10.118053$. Then we proceed the same way for L_2 and L_3, recording these values in a table like this.

LOWCAL	LOWCARB	LOWFAT
$\bar{x}_1 = 10.5$	$\bar{x}_2 = 7.5$	$\bar{x}_3 = 9$
$s_1 = 10.449880$	$s_2 = 7.501111$	$s_3 = 10.507140$

To calculate SS_w, we observe (see Exercise 3) that $\sum_{j=1}^{n}(x_{ij}-\bar{x}_i)^2 = (n-1)s_i^2$ and $n = 16$

so, $SS_w = (n - 1)[s_1^2 + s_2^2 + s_3^2] = 4138$

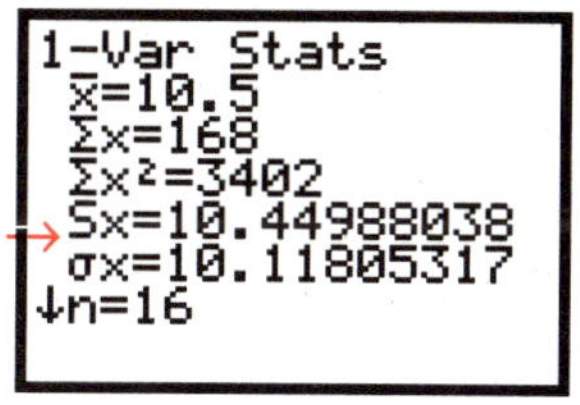

To calculate SS_b, we use equation ② to write $SS_b = n\sum_{i=1}^{3}(\bar{x}_i-\bar{x})^2$ where $\bar{x} = \frac{\bar{x}_1+\bar{x}_2+\bar{x}_3}{3} = 9$. Substituting for x_i from the table, and using $n = 16$, we obtain $SS_b = 16(1.5^2 + 1.5^2 + 0^2) = 72$.

Once SS_w and SS_b have been calculated, it remains only to evaluate the statistic $\frac{3(n-1)SS_b}{2SS_w} \approx 0.39149\ldots$

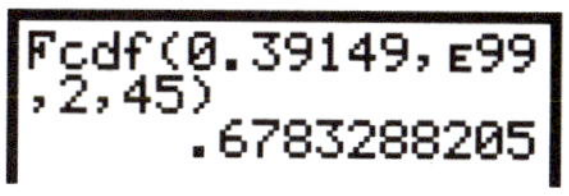

To find the probability that the F distribution with 2 and 45 degrees of freedom will take a value as large or larger than this, we enter the command shown in the display showing that the probability is about 0.678. That is, about two-thirds of the time when the null hypothesis is true, the F statistic will take a value as large or larger. We cannot reject the null hypothesis. In *Investigation* 8 p. 80, you will discover how to use the ANOVA test on your TI-83.

Copyright © 1998 by Brendan Kelly Publishing Inc.

Exercises

1. In the *Worked Example*, the ANOVA method revealed that the mean weight losses from each of the diets were not significantly different at the 5% (or even 30%) level. Does this mean that these diets are not significantly different in the weight losses they incur? Explain.

2. a) Write a sentence to explain the meaning of each phrase.
- error sum of the squares
- between sample sum of the squares
- total sum of the squares

b) Write a mathematical expression for each phrase in part a) and an equation that relates the expressions.

3. a) Given a sample $\{x_{i1}, x_{i2}, x_{i3}, \ldots, x_{in}\}$, with mean $\bar{x}_i$, write an expression for s_i^2, the mean square within the sample. (See Investigation 8 page 71.)

b) Use your expression in part a) to find an expression for $\sum_{j=1}^{n}(x_{ij}-\bar{x}_i)^2$ in terms of n and s_i.

4. a) Define L_2 as in the *Worked Example* on p. 79. Determine the values of $\bar{x}_2$ and s_2^2 using **1-Var Stats L₂**.

b) Use the expression you derived in Exercise 3b) to calculate $\sum_{j=1}^{16}(x_{2j}-\bar{x}_2)^2$.

5. A manufacturer offers three types of car tires: double-ply, triple-ply, and rugged radial. The company claims that the extra price for the premium brands is offset by longer lifetimes. The table shows the lifetimes of a sample of 10 tires of each type. Test at the 1% level of significance the null hypothesis that the mean lifetime is independent of the type of tire.

Tire Lifetimes in Thousands of Miles

Double Ply	30	46	53	67	48	52	28	31	37	42
Triple Ply	51	37	48	76	62	59	69	83	55	61
Radial	50	68	79	47	85	76	65	69	82	75

6. Determine the value that the ratio SS_b/SS_w in the *Worked Example* on p. 79 must exceed to cause a rejection of the null hypothesis at the

a) 5% level b) 1% level of significance.

(Hint: See Exercise 8 p. 62.)

7. a) Show that if x_{ij} are randomly selected values from a normal population with standard deviation σ, then

$$\frac{1}{\sigma^2}\sum_{i=1}^{3}\sum_{j=1}^{n}(x_{ij}-\bar{x})^2$$

has a chi-square distribution with $3n - 1$ degrees of freedom.

b) Show that SS_b/σ^2 has a chi-square distribution with 2 degrees of freedom and deduce that SS_w/σ^2 has a chi-square distribution with $3(n - 1)$ degrees of freedom.

Investigations

How do you perform ANOVA using the Test Menu on your TI-83?

8. In this investigation you will discover how to use the test menu on your TI-83 to perform an ANOVA without any computation!

a) Follow the keying sequence given in the *Worked Example* on page 79 to define the lists L_1, L_2, and L_3.

b) Press these keys to obtain the ANOVA command as shown in the display. [STAT] [▶] [▶] [ALPHA] [F]

```
ANOVA(L1,L2,L3)■
```

Then enter L_1, L_2, and L_3 as shown, and press [ENTER] to obtain this display.

```
One-way ANOVA
 F=.3914934751
 p=.6783265035
Factor
 df=2
 SS=72
 MS=36
Error
 df=45
 SS=4138
 MS=91.9555556
Sxp=9.58934594
```

Explain the meaning of each line in the display.
- F = .3914934751
- P = .6783265035
- Factor: df = 2, SS = 72, MS = 36
- Error: df = 45, SS = 4138, MS = 91.9555556
- Sx_p = 9.58934594

c) Use the ANOVA test on your TI-83 to test the hypothesis in Exercise 5. Use this to check your original answer.

9. a) The **augment(** command on your TI-83 list menu enables you to concatenate (combine) two lists. Combine lists L_1 and L_2 of the *Worked Example* into a single list L_4 by pressing these keys: [2nd] [LIST] [▶] [9]

Then enter L_1 and L_2 and store in L_4 as shown in the display. Next concatenate L_4 and L_3 and store in L_4. Use **1-Var Stats L₄** to find *SS* for L_4.

```
augment(L1,L2)→L4
```

b) Calculate SS_b as in the *Worked Example*. Use only your values of *SS* and SS_b to calculate SS_w. Verify the value given.

CHALLENGE

10. a) Prove the following identity.

$$(x_{ij}-\bar{x})^2 = (x_{ij}-\bar{x}_i)^2 + (\bar{x}_i-\bar{x})^2 + 2(x_{ij}-\bar{x}_i)(\bar{x}_i-\bar{x})$$

b) Use the identity in part a) to show $\sum_{i=1}^{3}\sum_{j=1}^{n}(x_{ij}-\bar{x})^2$

$$= \sum_{i=1}^{3}\sum_{j=1}^{n}(x_{ij}-\bar{x}_i)^2 + \sum_{i=1}^{3}\sum_{j=1}^{n}(\bar{x}_i-\bar{x})^2 + 2\sum_{i=1}^{3}\sum_{j=1}^{n}(x_{ij}-\bar{x}_i)(\bar{x}_i-\bar{x})$$

c) Show that the last term on the right side of the equation in part b) is zero and deduce that $SS_T = SS_w + SS_b$.

Copyright © 1998 by Brendan Kelly Publishing Inc.

Answers to Exercises & Hints for Investigations

Exploration 1

1. a) 4117 b) 112,154 c) 22.87% d) 6.53%

e) The data in the table is grouped. That is, we know how many people scored between 700 and 750, but we do not know how many scored exactly 700. Therefore, all we can say for certain is that the number of people scoring above 700 is between 9792 and 32,039.

2. b) With the cursor positioned on L_4 of the table, we enter: $L_2 + L_3$ and we obtain the table shown below. We then define Plot1 as shown in the display on the right below.

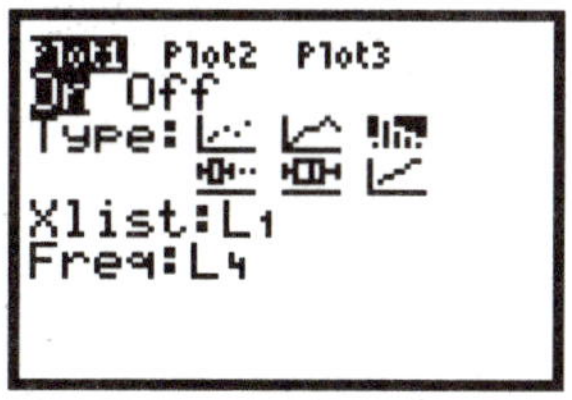

With the window settings shown on the display, we graph Plot1 to obtain the histogram shown here.

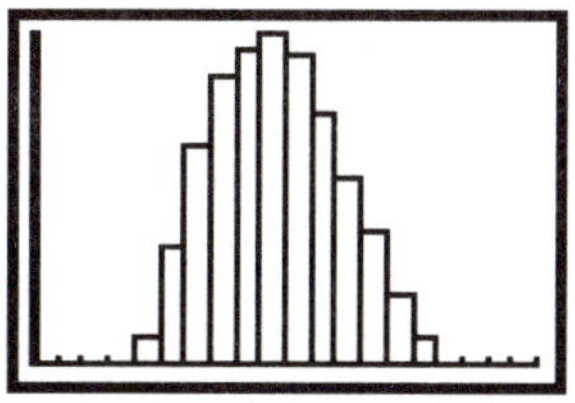

c) We obtain this display.

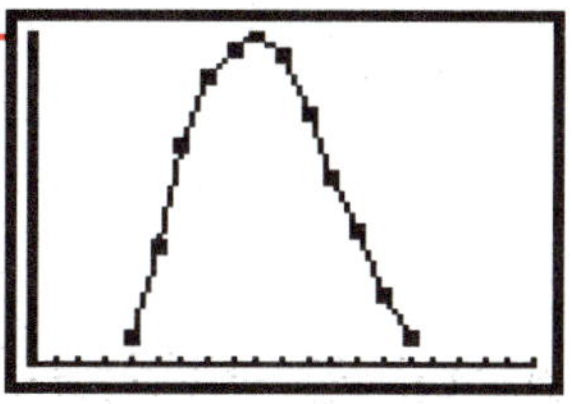

d) To change the class interval size to 100, we set Xscl to 100. Since we now have two intervals combined into one, we must increase the size of Xmax from 150000 to 300000. With the window settings on the left, we get the histogram shown on the right. By tracing along the curve to the interval $500 \le x < 600$, we discover that there are 254,176 students with scores in that interval.

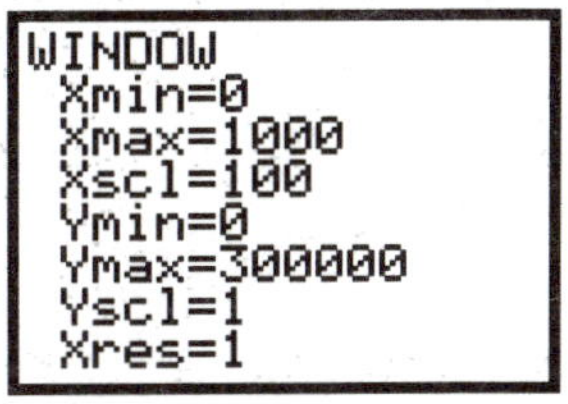

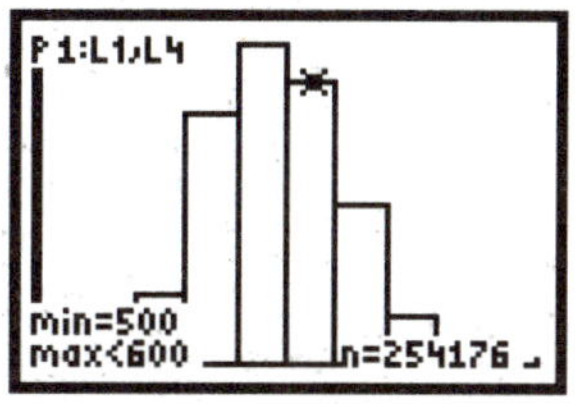

To find the total number of student scores, we evaluate **sum** L_4 using the LIST MATH submenu, and obtain the display 1025523. Therefore, 254,176 students out of 1,025,523 had scores in the 500s. That is about 24.8% of the students.

e) By tracing along the histogram to $x = 600$ and keeping a running total, or by creating a cumulative frequency function as in *Worked Example* 2, we find that there were 836,711 scores below 600. This is about 81.6% of all the students.

Exploration 1 (cont'd)

3. a) Your table should be the same as the one shown on page 11. We then define the plot and select the window settings as indicated in the displays below.

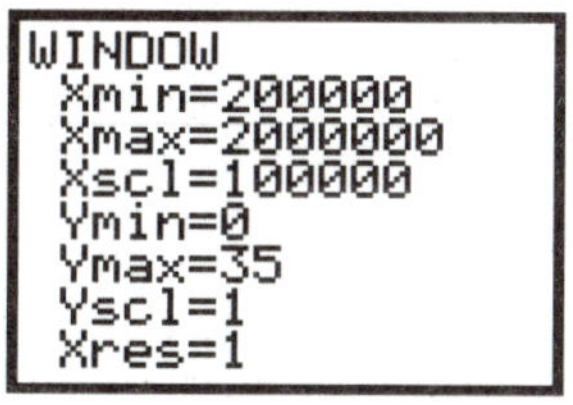

When we graph Plot1 with these settings, we obtain the histogram shown below left. We choose xyLine, to get the frequency polygon.

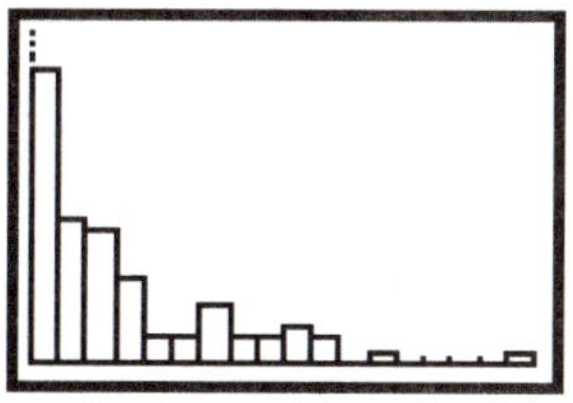

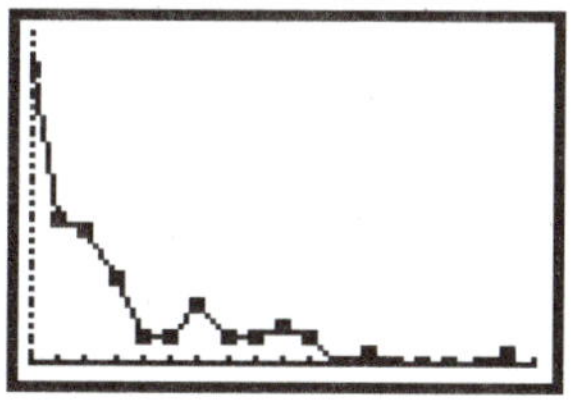

If we graph both plots simultaneously, we obtain this display. Observe that the frequency polygon joins the midpoints of the bars in the histogram.

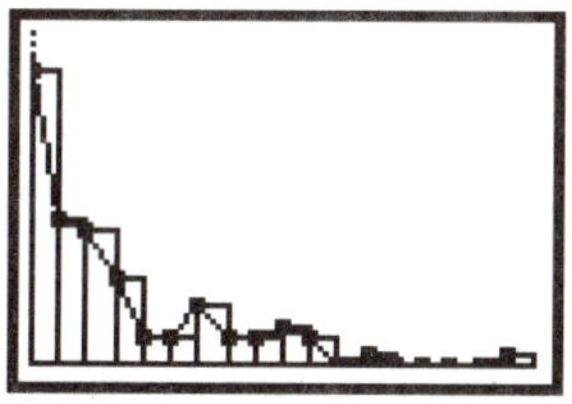

b) Adding the numbers in the table, we discover that there were 36 golfers out of the 96 golfers at or above $200,000 who won $500,000 or more. This is about 37.5% of the golfers over $200,000.

c) We set Xscl = 200000, Xmin = 100000 and Ymax = 35. We get the histogram shown here and trace to display 12 golfers or 12.5% in the interval $\$500,000 \le x < \$700,000$.

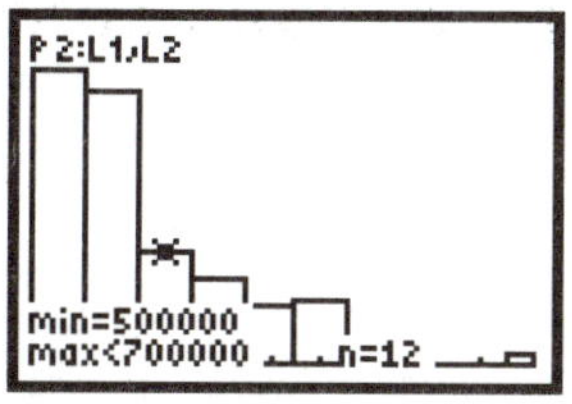

4. a) This display shows the ogive curve for the SAT scores of all students.

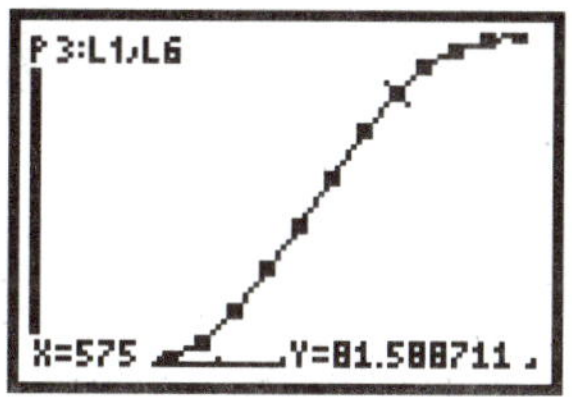

b) By tracing, we see that the cumulative percent of students whose scores were in the intervals up to the interval $550 \le x < 600$ (midpoint 575) was 81.58.

c) The ogive curve is steepest in the interval $450 \le x < 500$, because most of the student scores are in that interval, so the cumulative frequency is increasing the most in that interval.

Copyright © 1998 by Brendan Kelly Publishing Inc.

Answers to Exercises & Hints for Investigations

Exploration 1 (cont'd)

5. a) To change the definition of L_1, we define $L_1 + 100{,}000 \rightarrow L_1$. To form the cumulative sum L_3 we use the **cumSum(** command on the LIST menu. Then we define L_4 as $100L_3/96$. The ogive curve is shown here. When we trace along this curve to $x = 500000$, the display shows that Y = 62.5… That is, about 62.5% of the golfers are below $500,000 and 37.5% above. This agrees with exercise 3b).

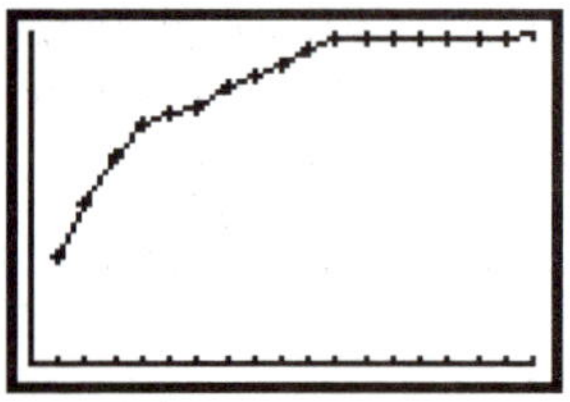

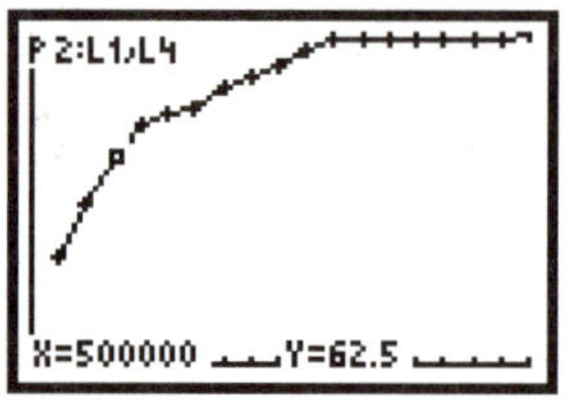

b) The curve is steadily increasing. (called *monotonic*) This is a property of all cumulative frequency curves because the height of the curve in any interval is the sum of all the heights up to and including that interval. Furthermore, the curve is very steep for small values of x and gradually flattens out as x increases. This indicates that there are relatively few huge money winners. These are generally the golfers who win several big tournaments in the same season.

Exploration 2

1. When there is a large set of data, we seek a single value around which most of the values cluster. The particular measure of central tendency that is most informative depends upon the kind of data we have and the kind of information we seek.

2. The median would not increase at all because there are 16 employees at $36,600 and so $36,600 will continue to be the middle salary. However, the mean salary takes into account the value of every salary and therefore, since $180,000 is greater than the previous mean its inclusion will increase the mean salary.

3. a) $\bar{x}$ = 26.416… , Q_1 = 16, Med = 29, Q_3 = 34
b) $\bar{x}$ = 18.545…, Q_1 = -17, Med = 42, Q_3 = 61

4. a & b) $\bar{x}$ = 403.3125, Med = 422

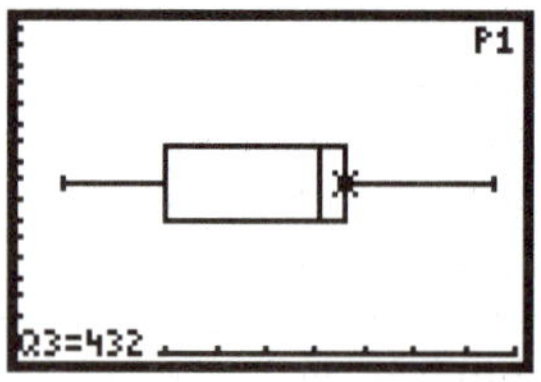

c) The quartiles are: Q_1 = 360, Med = 422, Q_3 = 432.
This can be verified by tracing along the box plot shown in the display.
The extremes are 318 and 492.
Alternatively, we could have calculated **1-Var Stats** and obtained the following displays.

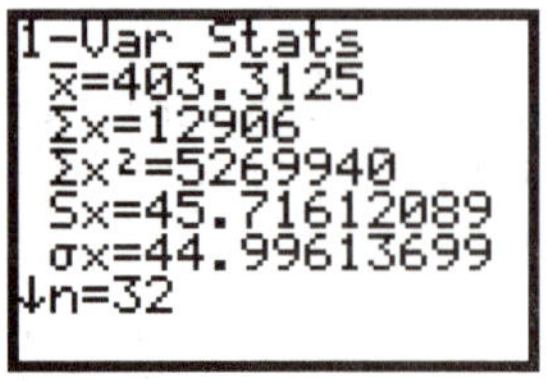

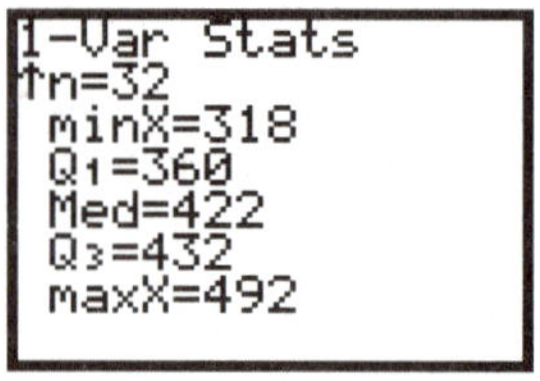

Exploration 2 (cont'd)

5. An *increase* of x% means that the new mean income is $(100 + x)$% of the original income. Therefore, the new mean income was: 0.8(172%) + 0.2(215%) or 180.6% of the original mean income. Therefore, the mean increase in income was 80.6% for the entire population.
An alternative method recognizes that since the non-whites represent 20% of the population, the average increase for the entire population will be the increase for the whites plus one-fifth the difference in the increase between the whites and non-whites. That is, 72% + 0.2(43%) or 80.6%.

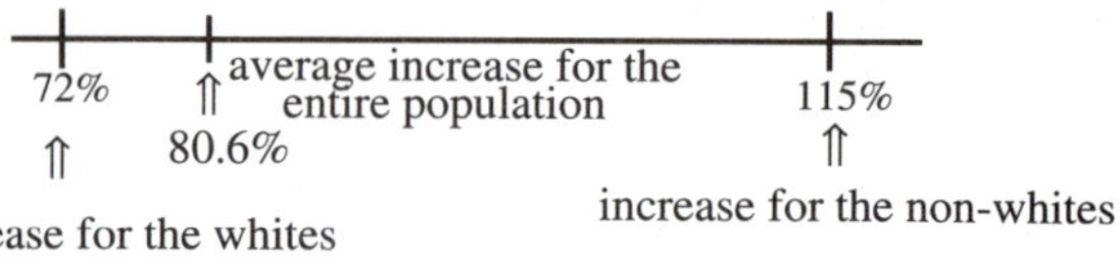

6. a) mean 50; median 50 b) mean 50.5; median $\frac{50+51}{2} = 50.5$
c) mean -1.5; median -1.5
Parts a, b and c suggest that the mean and median are the same for a set of numbers in arithmetic sequence. Is this true in general?

7.

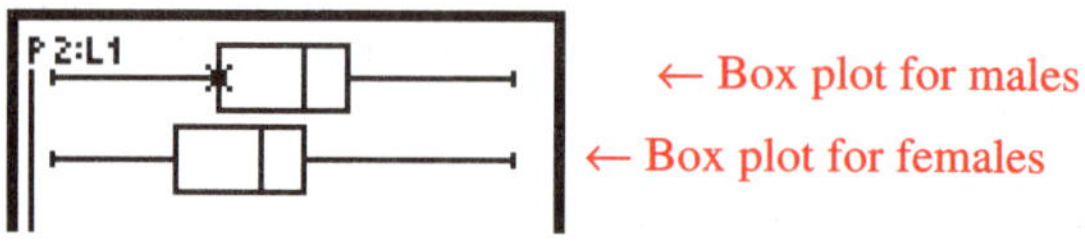

8. Using the stem-and-leaf plot to find the median, we observe that there are 30 measurements, so the median measurement is the mean of the 15th and 16th measurements when the numbers are arranged in ascending order. Counting along the leaves of the stem-and-leaf diagram to the 15th largest, reveals that the 15th and 16th lifespans are 13.2 and 13.3, so the median is 13.25. The first quartile is the middle number in the first 15 numbers, i.e. the 8th number, or 12.9. Similarly, the third quartile is the middle number in the last 15 numbers, i.e. the 23rd number or 13.6.

To obtain the mean from the stem and leaf plot, we multiply each stem by the number of leaves on it and then add to it the sum of the leaves. We obtain
$9 + 10 + (11 \times 2) + (12 \times 6) + (13 \times 15) + (14 \times 5) + 15.9 = 393.9$
Dividing the sum by 30, yields the mean of 13.13. ↑
Remember the leaves are tenths

To produce the box plot from the stem-and-leaf diagram, we enter the lifespans in L_1 of the STAT Edit menu.

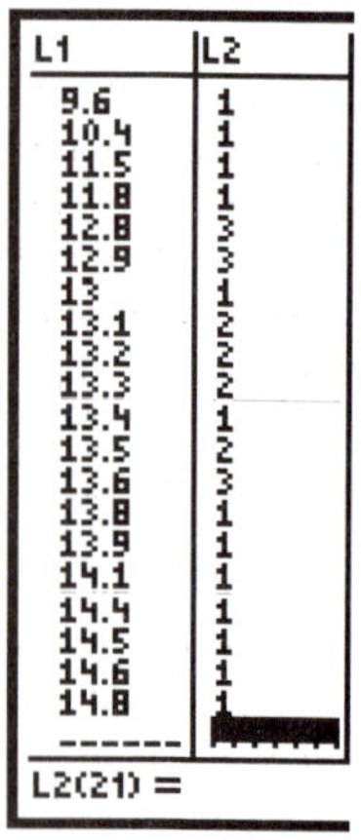

We enter the corresponding frequencies in L_2 as shown in the display. Then we define the Stat Plot by selecting these choices.

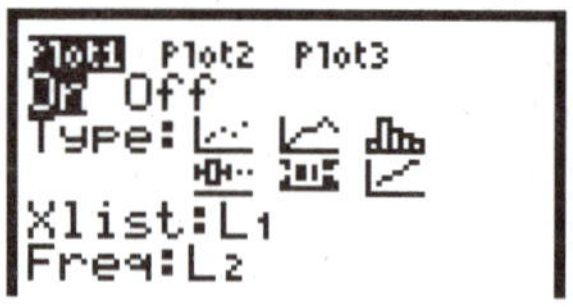

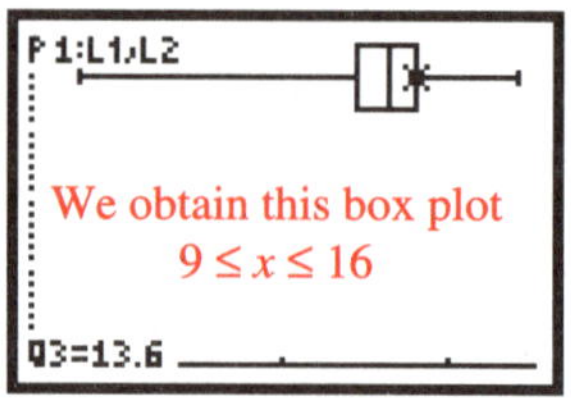

Copyright © 1998 by Brendan Kelly Publishing Inc.

Answers to Exercises & Hints for Investigations

Exploration 3

1. a) The median ages of the best actor and best actress award winners are respectively 43 and 34 years.

b) Since the leading man in a movie is usually paired romantically with the leading lady, and since males have been traditionally the older member of such couples, we would expect that the set of actors from which the "best actor" is selected is older than the set of actresses from which the "best actress" is selected. Recently the age difference between males and females in such roles may be declining. If so, we might expect this difference in the age gap to be less in the second half of this quarter century.
A variety of other answers are also feasible.

c) The age range of the "best actors" is from 30 to 76 years.
The age range of the "best actresses" is from 21 to 80 years.

d) For reasons given in part b), the set of actresses from which the "best actress" is selected is younger than the set of actors from which the "best actor" is selected. Hence, it is more likely that the youngest best actress will be younger than the youngest best actor. However, since it takes only one person to make a new minimum, the behavior of the extreme values is much less predictable than the behavior of the median.

e) Actors and actresses who reach advanced age are less likely to be starring as couples in a romantic involvement so it would seem that the argument expressed in part b) does not apply to this age range. Another factor is the fact that women live longer than men. There would be more actresses in the advanced ages i.e. 70's, 80's and 90's than actors in these age ranges. Hence, the pool from which "best actresses" are selected might be expected to have more candidates of advanced years.

f) Together, the arguments expressed in parts b) and e) suggest that the pool from which "best actresses" are selected has greater numbers at both ends of the age scale. It would seem therefore that the "best actresses" should have a larger range than the "best actors". In fact the "best actresses" have an age range of 80-21 or 59 years. The "best actors" have an age range of 76 - 30 or 46 years.

g) The interquartile range for "best actresses" is 41 - 29 or 12 years. The interquartile range for "best actors" is 49 - 39 or 10 years.

h) The median age for winners of best actress is 9 years younger than the median age for winners of the best actor award. If there are more actresses at the advanced ages, these large values would not affect the median, but would increase the mean age of the actresses, thereby decreasing the gap in the median ages. When we use OneVar Stats, we find the mean age for "best actresses" is 36.96 years. The mean age for "best actors" is 44.76 years. The difference in means is about 7.8 years.

2. a) 50% b) 100%
c) When the median is the mean of the first and third quartiles.
d) Half of the values cluster closely near the median and in the remaining values, there is at least one small and one large outlier.

Exploration 4

1. Measures of dispersion were developed to determine how tightly a set of data cluster around various measures of central tendency.

2. a) Range in lifetimes for Brand A batteries: 9.3 – 3.1 or 6.2 years
b) Range in lifetimes for Brand B batteries: 7.5 – 3.3 or 4.2 years
c) The range takes into consideration only the difference between the largest and the smallest values in a set of data and ignores all the other values. Therefore, two sets of data may have the same range but significantly different clustering around the mean.

3. a) The worked example indicates that the mean is 57 years.
The mean deviation is 31/30 or 1.033… .

b) The mean deviation for Brand B batteries is 17/30 or 0.566…

c) Since Brand B batteries have a smaller mean deviation in their lifetimes, the mean life of 5.7 years is a more reliable measure for Brand B than for Brand A batteries.

4. a) $\bar{x} = 6$; $s = 3.316...$ b) $\bar{x} = 0$; $s = 3.316...$
The standard deviation of a set of data does not change when a fixed constant is added to all the values. In essence, adding a constant to all the values merely *slides* the data points horizontally without increasing their distances from the mean.

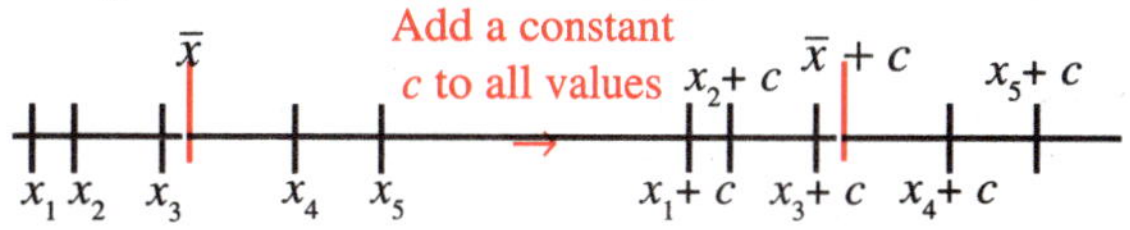

5. For the set of data, S = {1, 2, 3, 4, 5}, the mean deviation is 1.2. The standard deviation is $\sqrt{5/2}$. We add the data point, 81 to S, to obtain T = {1, 2, 3, 4, 5, 81}. The mean deviation of T is $\frac{65}{3}$.
The standard deviation of T is 31.874...
For the set, S, the standard deviation is about 0.93 times the mean deviation, but for T the standard deviation is about 1.47 times the mean deviation. That is, the standard deviation seems more sensitive to extreme values.

6. Diameters of 30 ball bearings from Acme:

$$\bar{x} = 8.0163;\ s = 0.1038...$$

Diameters of 30 ball bearings from Delta: $\bar{x} = 8.0296$; $s = 0.0536...$
Though the mean diameter for the ball bearings from Acme is slightly closer to 8 mm than those from Delta, the standard deviation for the Delta ball bearings is about half that for Acme. That is, the Delta ball bearings have smaller deviations from the mean. Since the difference in the means is negligible relative to the difference in the standard deviations, and since most of the diameters will be within a standard deviation of the mean, the ball bearings from Delta have smaller deviations from 8 mm.

7. Histogram a) is the one which corresponds to the lifetimes of Brand B batteries. We can see this by rotating the double stem-and-leaf plot on page 19 by 90 degrees.

8. When the dispersion of a set of data is small, most of the area under the bars is concentrated near the mean value as in histogram a) in exercise 7. For a set of data of greater dispersion, the area under the bars is distributed further away from the mean as in histogram b).

Copyright © 1998 by Brendan Kelly Publishing Inc.

Answers to Exercises & Hints for Investigations

Exploration 5

1. In general, two variables are said to be *positively correlated* if large values of each variable are associated with large values of the other variable. The degree of this correlation may be described as weak or strong depending upon how few violations of this relationship exist in the set of data points.
Two variables are said to be *negatively correlated* if large values of each variable are associated with small values of the other variable. When the correlation between two variables is approximately linear, the data points in the corresponding scatterplot cluster along a straight line and most of the data points can be contained within an eccentric ellipse with the straight line as its major axis.
When two variables are not correlated, we say that they show *no correlation* (or that they are *uncorrelated*).

2. a) strong negative correlation b) positive correlation
c) no correlation d) negative correlation

3. The scatterplot for the ages of women and their spouses is shown below. Since the data points cluster along a straight line of positive slope, we say the ages of women are positively correlated to the ages of their spouses. Since the clustering is close to a line, we would describe this relationship as a *strong positive correlation.*

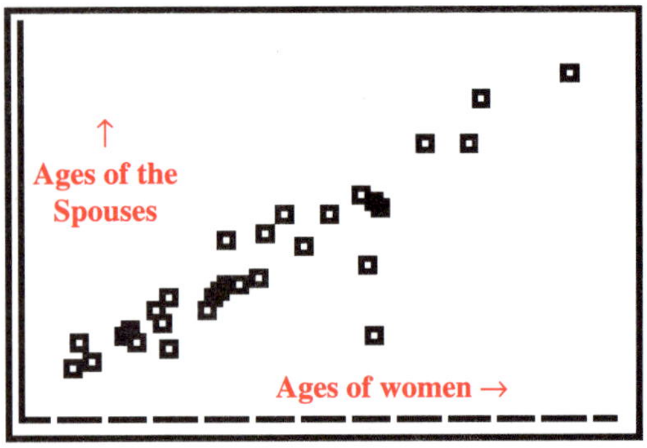

4. The scatterplot for the number of cricket chirps per minute and the outside temperature is shown below.

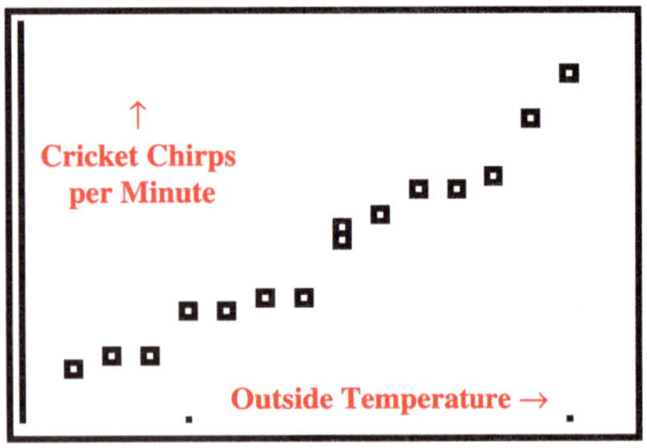

This scatterplot shows that there is a strong positive correlation between the number of cricket chirps per minute and the outside temperature. The scientist's hypothesis seems to be valid. Some might suggest that advancing such a hypothesis without using a scatterplot wouldn't be cricket. We are inclined to chirp in agreement.

Exploration 5 (cont'd)

5. a) A positive correlation: Taller people will tend to be heavier than shorter people, so a large height is generally associated with a large weight.

b) A positive correlation: Substantial personal savings are usually associated with increased age since few young people have had the opportunity to accumulate wealth.

c) A positive correlation: The largest incomes are usually earned by people who are highly educated and the lowest incomes by people who have little or no formal education.

d) A negative correlation. Cities of large latitude are closer to the earth's poles than cities of smaller latitude and are therefore colder (lower mean temperature) in winter than cities of smaller latitude.

e) No correlation. For an individual, golf score may drop as the person improves during middle age. In older age this trend may reverse itself. However, if a data point consists of a randomly selected person's age and his (or her) golf score, then there would be little or no correlation expected, since golf scores would seem to be randomly distributed at every age level depending on experience and skill.

f) A strong negative correlation. Travel time for a particular trip is inversely proportional to the speed of travel. Since large travel times correspond to small travel speeds, these variables are negatively correlated. Since there are *no* deviations from this correspondence, the negative correlation is strong.

g) No correlation. Unless the olives are consumed with dry martinis, there is little reason to expect any correlation between these variables.

6. a) If $y = ax + b$, then $y_i = ax_i + b$ and $\bar{y} = a\bar{x} + b$, so $y_i - \bar{y} = a(x_i - \bar{x})$. Substituting into the formula for r, we obtain: $r = \frac{a}{|a|}$ and so, $|r| = 1$. That is, $r = 1$ when $y = ax + b$ and $a > 0$. $r = -1$ when $y = ax + b$ and $a < 0$.

b) Since there is no universally agreed range of r corresponding to the terms *strong correlation* and *weak correlation*, we show the general ranges on a spectrum.

$r = -1$ $r = 0$ $r = 1$

strong negative correlation — weak — no correlation — weak — positive correlation

The (linear) correlation coefficient between the number of cricket chirps per minute and the outside temperature is: $r = 0.9768\ldots$

7. a) The (linear) correlation coefficient between the age of a woman and the age of her spouse is: $r = 0.90716\ldots.$
b) The (linear) correlation coefficient between the hours of study and the test mark is $r = 0.7650\ldots.$
Note: Compare these coefficients with the correlation of 0.4 between the IQ's of parents and their children, or the correlation of 0.5 between the intelligence of siblings.

Copyright © 1998 by Brendan Kelly Publishing Inc.

Answers to Exercises & Hints for Investigations

Exploration 6

1. Definitions are given on page 25.

2. Following the step-by-step instructions, we divide the measurements into three groups: 1928-1952; 1956-1972 and 1976-1996. The summary points for these groups are respectively, (1936, 10.30), (1964, 10.00) and (1986, 9.95). We join the first and third summary points with a line segment and then draw a parallel line through the middle summary point. We then slide a ruler one-third the distance from the line segment to the line and trace the median-median line. To do this accurately, extend the line segment to where it cuts the y-axis and then slide one-third the distance up the y-axis to the y-intercept of the line.

Men's Winning Times in the 100-*m* Run

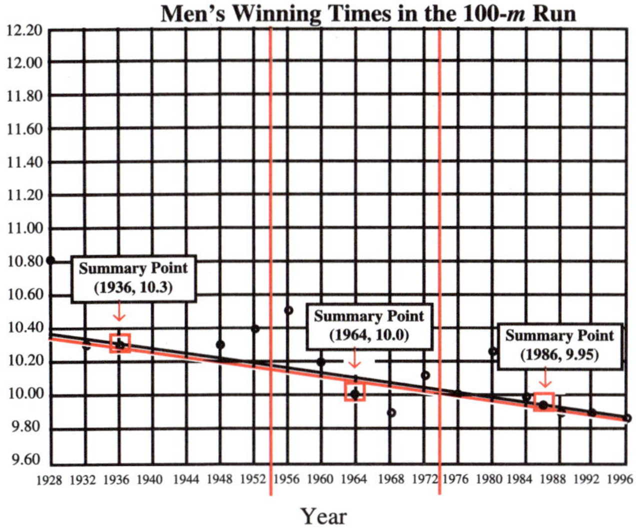

When we construct the median-median line on our calculator as in the worked example, we obtain the display shown below. Using the STAT CALC menu, we obtain the equation of the median-median line indicating that its slope is -0.006 and y-intercept 22.810. Note that the slope of the median-median line is the same as the slope of the line segment joining the first and third summary points. Therefore, we can calculate the slope of the median-median line by calculating the slope of the line segment.

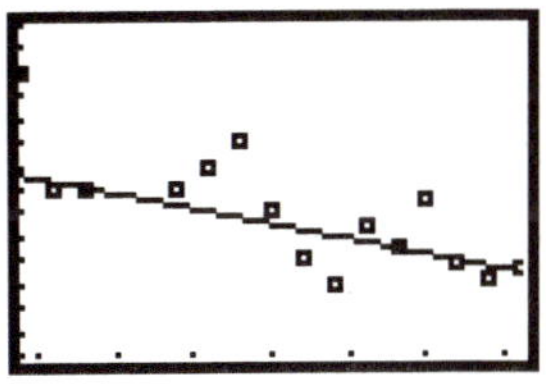

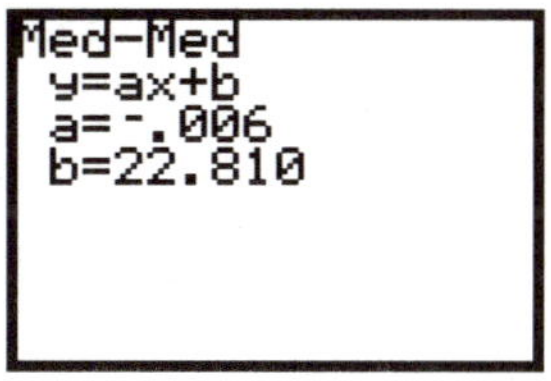

Note: The median-median line is an effective device for smoothing out fluctuations in data and looking for long term trends, in such applications as sale analysis, stock market indices and monetary exchange rates. However, caution must be exercised in making predictions. The median-median line approach is based on a linearity assumption. If this assumption is incorrect (and it usually is incorrect) the predictions have no validity.

Exploration 7

1. The method of least squares is a method for finding the equation of a particular curve in a family of curves (such as linear, logarithmic, exponential etc.) for which the squares of the vertical distances of the data points from that curve is a minimum. This curve is called the *regression curve.* When seeking a straight line (called the linear model), we choose the straight line of best fit and it is called the *regression line of y on x.* However, if we are seeking an exponential or logarithmic curve to approximate our data, the line of best fit is referred to respectively as the *exponential* or the *logarithmic* regression curve of y on *x*.

The residual of a data point is the observed value of the dependent variable minus its predicted value (obtained by substituting the value of the independent variable into the equation of the regression curve).

2. If both axes have the same scale then the slopes are as follows:
a) slope ≈ – 0.8 b) slope ≈ 0.8 c) slope ≈ – 0.6

3. Both the median-median line and the regression line are straight lines which attempt to describe a linear relationship between the independent and dependent variables. They both assume an underlying linear relationship between the two variables and are used to smooth out data and predict values of one variable from the other.

The differences between a median-median line and a regression line are similar to the differences between a median and a mean. Substantial increases or decreases in the extreme values of a set of data will have no effect on the median-median line but it will affect substantially the equation of the regression line. That is, outliers have minimal effect on a median-median line but play a significant role in determining a regression line.

4. The curve of best fit:
- in the logarithmic model, has equation $y = a + b\ln x$
- in the exponential model, has equation $y = ab^x$
- in the power model, has equation $y = ax^b$

where "a" and "b" are parameters.

5. To compare regression curves of various types, we find the regression equation for various models, such as; linear, logarithmic, exponential etc. and calculate the residuals as in the *Worked Example* on p. 29. The curve for which the sum of the squares of the residuals is smallest is the curve we select as the regression curve.

6. a) If the relationship were linear, increasing the hours of study by one hour should increase the mark by a fixed amount no matter what mark a student has. However, a student with a high mark say, 99, cannot improve the exam mark substantially, no matter how many hours she works. This is called *the law of diminishing returns*.

b) The law of diminishing returns suggests a logarithmic relationship. The opportunity to improve a mark is greatest when the study hours are small. Once a student has invested large numbers of hours in study, the mark is so high that there is little room for improvement; i.e. increasing *x* by a lot, generates small growth in y.

Copyright © 1998 by Brendan Kelly Publishing Inc.

Answers to Exercises & Hints for Investigations

Exploration 7 (cont'd)

7. a & b) The scatterplot is shown below along with the graphs and equations of the linear and logarithmic regression lines.

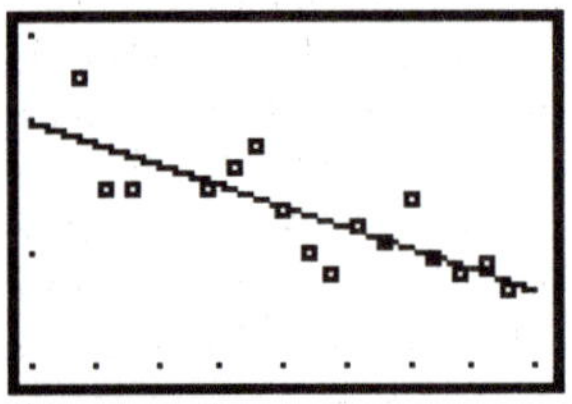

Observe that only one regression line is visible because they are so close that they are almost coincident.

To examine each line in turn, merely deselect one of the curves on the **Y=** list.

To determine which regression line gives a better fit, we calculate the sums of the squares of the residuals for both curves, following the procedures demonstrated in the *Worked Example*, page 29. The sum of the squares of the residuals for these two regression lines are 0.3505 and 0.3499 for the linear and logarithmic lines respectively. It seems that the logarithmic model is a slightly closer match.

c) When we evaluate substitute x = 2060 into the logarithmic regression line, we obtain a prediction of about 9.29 seconds. The median-median line predicted a winning time of 9.48 seconds.

Note: This display shows in order from top to bottom the exponential, logarithmic and linear regression lines in the interval: $2035 \le x \le 2070$, $9.13 \le y \le 9.60$

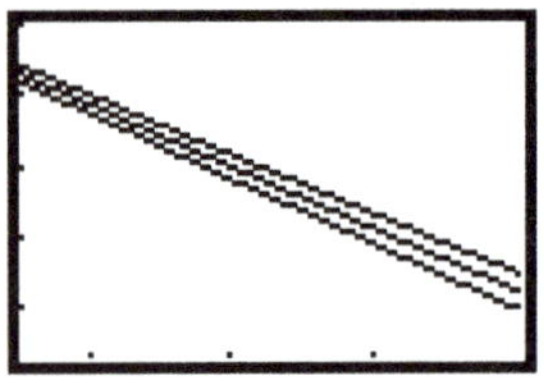

Since the linear model predicts a 0 winning time eventually, it is not reliable too far outside the data range.

8. The scatterplots and two regression lines are given in the display.

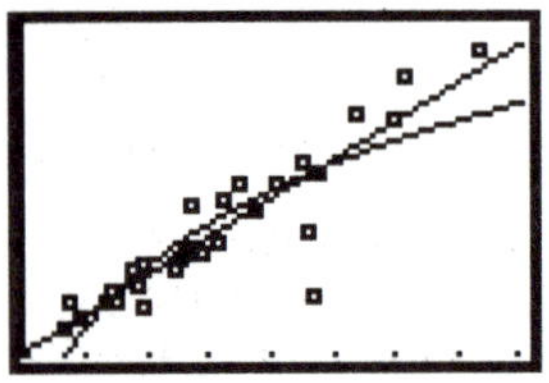

a) r = 0.90716... This is a strong correlation.
Equation of linear regression line is: Y= 2.811...+ 0.94869...X

b) r = 0.8699...The linear regression line provides a better match to the data because the sums of the squares of the residuals is smaller than for the logarithmic regression line.
Equation of logarithmic line is: Y = -90.264...+ 36.421...lnX

If men are generally x years older than their wives at marriage, then Man's age = Spouse's age + x. That is, we have a linear relationship. At advanced ages women's greater longevity could affect this unless the age of dead spouse is contained in the data.

c) The estimated age of the husband of a 48-year-old using linear regression is about 48.34 years.

Exploration 8

1. The theoretical probability of an event is the result of a computation based on the assumption that a certain set of outcomes are equilikely. The experimental probability of an event is an estimate of its theoretical probability based on the observed relative frequency of that event on a large number of replications.

2. Since there is no reasonable way to compute the theoretical probability of each outcome, we compute the experimental probability by tossing the cup a large number of times and using the relative frequency of each outcome as the experimental probability.

3. a) The U.S. Department of Health and Human Services has reported that in 1989, about 934,000 Americans died from heart disease. This represented approximately 43% of all American deaths in that year. We take this as an estimate of the relative frequency of deaths from heart disease that will occur in future.

b) This probability is based on observation and is therefore an experimental probability. Causes of death are too complex to admit a simple model that could yield a theoretical probability.

Since the answers to the investigations which follow are based upon simulations, they will vary. Consequently we provide theoretical probabilities where known, so that you can compare your results to what is predicted. If your answers differ significantly from the probability given, repeat your simulation for more replications.

4. a) The theoretical probability of heads is 0.5, so the expected frequency is about 50. However, the theoretical probability of *exactly* 50 heads is: ${}_{100}C_{50}/2^{10}$ or 0.0795, so it is not likely you will get exactly 50 heads. However, the probability that the number of heads, H, satisfies $45 \le H \le 55$ is about 0.73. Therefore, the frequency of heads should be 50 ± 5 about 3 times out of 4.

b) On each of the 10 replications of the simulation, it is just as likely that the frequency of heads will fall short of 50 as exceed 50. Furthermore, from part a, we expect that most of the 10 frequencies will fall between 45 and 55, so the median should be close to 50.

5. Answers will vary somewhat, but the relative frequency of heads will *usually* approach closer to one-half as the number of tosses increases. In a small proportion of simulations there will be exactly 25 heads on 50 tosses, so it will not be possible in such cases that 1000 tosses will yield a relative frequency closer to one-half. However, since the dispersion of the set of relative frequencies decreases as the number of tosses increases, we can expect the relative frequency of heads to approach closer to one-half as the number of tosses increases.

6. a) The probability that A will win is 3/4, so we would expect that A will win about 75 of the 100 games. The probability that A will win *exactly* 75 of the games is about 0.0918; however, the probability that A will win between 70 and 80 games ($70 \le x \le 80$) is about 0.80.

b) The median number of wins for A should be close to 75.

c) The results in part b will probably indicate that the median number of wins for A is closer to 75 than to 66, so the Pascal model seems to fit observations more closely than the Roberval model.

7. Most people expect the number of heads and tails to balance and they expect H - T to approach 0. However, theoretical probability that H - T will be exactly 0 after 100 tosses is about 0.0795. It is the expression (H – T)/N that approaches 0 as N becomes very large.

9. a) 0.5 b) 0.33... c) 0.833... d) 0.33...

Copyright © 1998 by Brendan Kelly Publishing Inc.

Answers to Exercises & Hints for Investigations

Exploration 9

1. A simulation is a method for obtaining the experimental probabilities for various outcomes of an experiment by running a large number of replications of a similar statistical experiment whose outcomes should have the same probabilities as those we wish to measure. By observing the relative frequencies of the outcomes in the simulation, we can estimate the probabilities we seek to measure.

2. There is no way we can ever be sure that an experiment is behaving in a typical way, and so we never know when an experimental probability is a close approximation to the corresponding theoretical probability. But, as we increase the number of replications of the simulation, we reduce the chances that such an aberration will occur.

3. Since a correct and an incorrect guess on an item are equilikely, we can use **cointoss** to simulate guessing. We can associate with the outcome, "heads", the outcome of selecting a correct answer.

a) The theoretical probability of exactly 7 correct responses is ${}_{10}C_7/2^{10}$ or 0.117.

b) The theoretical probability of at least 7 correct is 0.1718… Our simulation should produce at least 7 heads on about 17% of the simulations and this estimate should become more reliable as the number of replications is increased.

4. Since both genders are equally likely at birth, we can use the **cointoss** simulation and identify "heads" with the occurrence of a girl.

a) The theoretical probability of exactly 4 girls is ${}_6C_4/2^6$ or 0.234..

b) The theoretical probability of at least 4 girls is:

$({}_6C_4 + {}_6C_5 + {}_6C_6) / 2^6$ or 0.343…

The **cointoss** simulation should yield at least 4 heads on about one-third of the replications.

5. Since each item offers 6 equally likely choices, we can simulate the selection of an answer with the toss of a die. We identify rolling a one with the selection of a correct answer. The process of selecting 20 answers can be simulated by running **dietoss** for 20 tosses. The theoretical probability that the student will *not* choose correctly on more than one-quarter of the items, is about 0.898…. That is, the theoretical probability that the student will choose the correct answer on at least 6 of the items is 0.1018…

The **dietoss** simulation should yield more than 5 correct responses on about 10% of the replications.

6. The probability that a female is hired for a particular job in that occupation is one-third. Therefore, we can simulate the hiring of 82 employees for Acme by 82 tosses of a die. We identify the occurrence of a 1 or a 2 on the die with the hiring of a female and any other outcome with the hiring of a male. We run **dietoss** with N = 82 and record the total number of tosses on which a 1 or 2 occurs. Repeat this process about 20 times The theoretical probability that 21 or fewer females would be hired as a result of random selection, is about 0.0837…. That is, between one and two replications out of 20 should yield a total of 21 or less.

This analysis is not enough to prove discrimination, since all of the employees may have been hired at a time when the percentage of women in the work force was much smaller. Many factors would need to be considered before inferences could be drawn.

Exploration 9 (cont'd)

7. Since there is one chance in 6 that a passenger will not show, we can simulate the 96 decisions of the booked passengers by 96 tosses of a die and we can identify the outcome "one" (i.e. one dot) with the decision not to show. We run DIETOSS simulation for 96 tosses and record whether the outcome "one" occurs fewer than 11 times, creating a seat shortage. We repeat this simulation a total of 20 times and record the fraction of simulations on which there are fewer than 11 occurrences of "one". This gives us an estimate of the probability that more than 85 of the 96 booked passengers will show. The theoretical probability that more than 85 passengers will show is:

$${}_{96}C_0\left(\frac{5}{6}\right)^{96} + {}_{96}C_1\left(\frac{1}{6}\right)^{1}\left(\frac{5}{6}\right)^{95} + {}_{96}C_2\left(\frac{1}{6}\right)^{2}\left(\frac{5}{6}\right)^{94} + \ldots + {}_{96}C_{10}\left(\frac{1}{6}\right)^{10}\left(\frac{5}{6}\right)^{86} \approx 0.06$$

That is, there is only about a 6% chance that there will be a seat shortage.

8. This problem can be simulated with a single die since there are only 6 positions. We run the simulation **dietoss**, for 20 tosses and record whether there is at least one zero in the list displayed. If not, all 6 outcomes have occurred at least once. That is, the player has been in all 6 positions at least once. We call this the "non-zero" outcome. To estimate the probability, we repeat the simulation 30 times and record the relative frequency of the non-zero outcome over the 30 replications.

The theoretical probability that a particular member of the Old Timers will have played all positions after 20 games is about 0.85.

9. b) The actual areas are:

(i) $\pi\sqrt{40}$ (ii) $3\sqrt{3}\pi$ (iii) $\dfrac{97\pi}{\sqrt{84}}$

10. a) When we ran ELLIPSE with $A = \sqrt{2}$ and $B = \sqrt{7}$ and N=1000, we obtained these displays: (Remember:press ENTER when the points have finished plotting.)

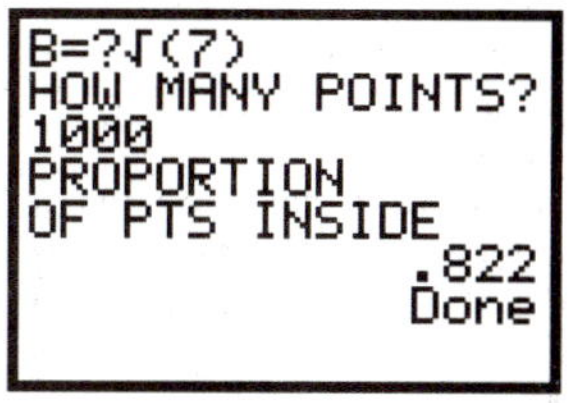

Estimated area 0.822(4√14)=3.288√14. The formula for the area inside an ellipse of semi-major and semi-minor axes *A* and *B* (see *Exercise* 9b) is πAB, so we see that the true area is 3.141√14 and the error in the estimated area is less than 5%.

When we ran ELLIPSE with $A = \sqrt{2}$ and $B = \sqrt{7}$, and N = 5000, we obtained these displays.

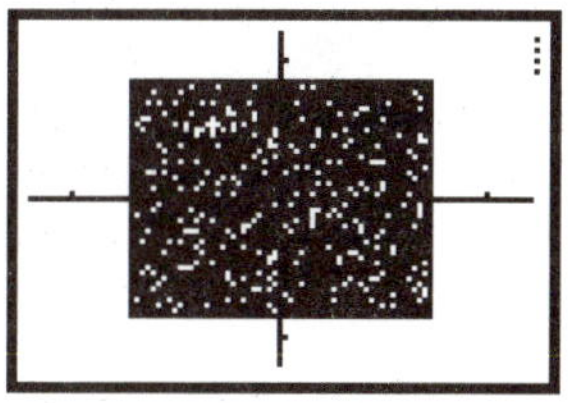

Note that the greater number of replications yielded the closer approximation 0.775(4√14) = 3.10√14 to the true area.

Copyright © 1998 by Brendan Kelly Publishing Inc.

Exploration 9 (cont'd)

10. b) The program, HCYCLOID is shown here.

```
PROGRAM:HCYCLOID
:Prompt C
:C^(3/2)→A
:0→Xmin
:A+1→Xmax
:0→Ymin
:A+1→Ymax
:Input "HOW MANY
 POINTS? ",N
:0→S
:ZSquare
:"(C-X^(2/3))^(3
/2)"→Y1
:DispGraph
:Line(0,A,A,A)
:Line(A,A,A,0)
:For(I,1,N,1)
:Arand→X
:Arand→Y
:Pt-On(X,Y)
:If X^(2/3)+Y^(2
/3)<C
:S+1→S
:End
:S/N→T
:Pause
:Disp "PROPORTIO
N"
:Disp "OF PTS IN
SIDE",T
:
```

c) When we ran HCYCLOID for 100 points, we got these displays.

```
C=?3
HOW MANY POINTS?
100
PROPORTION
OF PTS INSIDE
            .320
            Done
```

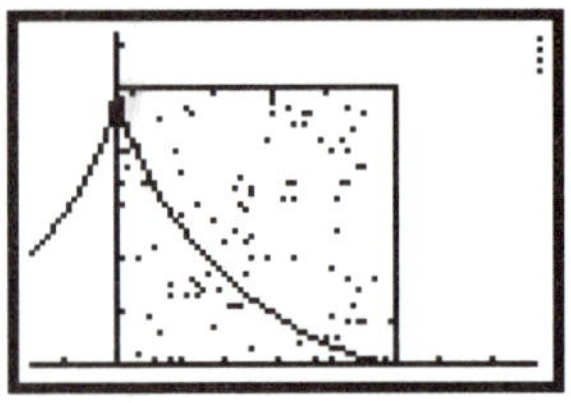

Formula for area under a hypocycloid with $c = 3$ is: $(3/32)\pi c^3$ or 7.95….

The simulation indicates that the area under the hypocycloid is 32% of the area of the square. But the square has side length $3\sqrt{3}$ and therefore area 27 square units. The area under the hypocycloid is therefore approximately 32% of 27or 8.64 square units. The estimate is within 10% of the exact value.

d) The area is given by the formula $A = k\pi c^3$. On substituting $A = 8.64$ from part b) and solving for $k = A/27\pi$, we get $k = 0.102$.
The actual value for k is 3/32 or 0.0937…, so our estimate of k is within 10% of the exact value. If we increased our simulation to 1000 points we would most likely obtain an estimated value of k even closer to the true value.

11. a) When we ran BINOMIAL for 10,000 tosses, we obtained the histogram in the display showing the frequency of each of the 8 possible outcomes; i.e. 0 boys, 1 boy, 2 boys……7 boys. The table shows the frequency (L_2) of each outcome (L_1) and the relative frequency (L_3) of each outcome.

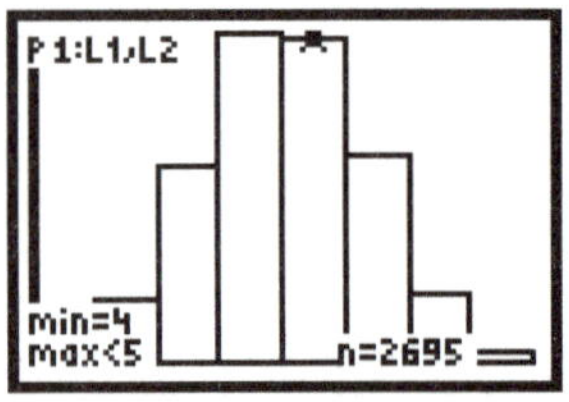

L1	L2	L3 2
0	78	.0078
1	541	.0541
2	1630	.163
3	2721	.2721
4	2695	.2695
5	1693	.1693
6	576	.0576
7	66	.0066
------		------
L2(9) =		

b) The theoretical probability of exactly 4 boys out of 7 children is ${}_7C_4 \times 2^{-7}$ or ≈ 0.273. Running the BINOMIAL simulation, we obtained (See table above) an estimate of $0.2695 \approx 0.270$. This shows close agreement between theoretical and experimental probabilities.

12. a) & b) See *Worked Example* 1 on page 40. The theoretical probability is very close to zero.

Exploration 10

3. The theoretical probability is given by:

$1-[0.8^{20}+20(0.2)(0.8)^{19}+{}_{20}C_2(0.2)^2(0.8)^{18}+{}_{20}C_3(0.2)^3(0.8)^{17}$
$= 0.58855…$

4. The theoretical probability of 10 or more snake eyes $\approx$ 0.002.

5. In factorial notation, the expression is: $\dfrac{364!}{342!(365)^{22}}$

We can evaluate the quotient 364!/342! by evaluating ${}_{364}C_{342}$ and then multiplying the result by 22!. We then divide by the power 365^{22}, and the display shows 0.493… However, if we tried to evaluate 364! directly, we would get an overflow error.

6. b) The theoretical probability is: $1-\left(\frac{11}{12}\right)\left(\frac{10}{12}\right)\left(\frac{9}{12}\right)\left(\frac{8}{12}\right)$

which is equal to 0.618…

8. Program POLYHEDR shown here simulates the situation. The time X of Polyhedron's departure is selected randomly from a 15-minute interval and the time Y of the paper arrival is selected randomly from a 20-minute interval so they have a 5-minute overlap. This is run 5 times (5 days). The relative frequency of success, i.e. Y occurs before X, is displayed. In the answer to exercise 8 of the next *Exploration*, the theoretical probability is shown to be 0.1916.

```
PROGRAM:POLYHEDR
:Input "HOW MANY
 TRIALS?",N
:0→L
:For(I,1,N)
:0→K
:For(J,1,5)
:15rand→X
:10+20rand→Y
:If X-Y≤5 and X≥
Y
:K+1→K
:End
:If K>0
:L+1→L
:End
:Disp "RELATIVE
FREQ OF SUCCESS"
:Disp L/N
:
```

10. b) The probability of coincident birthdays for each given n are:

i) 0.117	ii) 0.253	iii) 0.411
iv) 0.569	v) 0.706	vi) 0.814

12. When we ran BINOMIAL for 10000 tosses of 10 coins, we obtained the histogram shown in this display

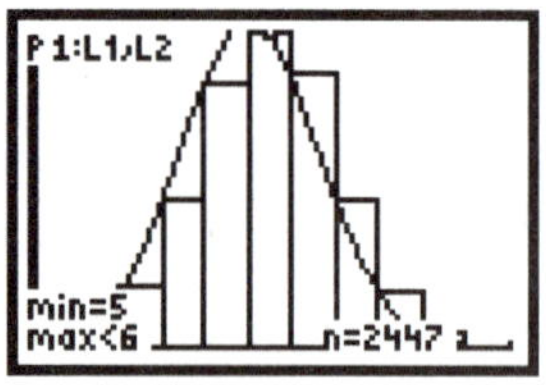

Superimposed curve has equation $y= 10000/(\sqrt{(5\pi)})e^{\wedge}(-(x-5)^2/5)$
Observe that the curve passes near the upper left corner of most bars.

When we ran BINOMIAL for 10000 tosses of 20 coins, we obtained the following table of frequencies and relative frequencies for the 21 possible number of heads 0, 1, … 20.

L1	L2	L3 2
0	0	0
1	1	1E-4
2	2	2E-4
3	9	9E-4
4	55	.0055
5	138	.0138
6	370	.037
7	744	.0744
8	1162	.1162
9	1572	.1572
10	1697	.1697
11	1654	.1654
12	1242	.1242
13	783	.0783
14	385	.0385
15	130	.013
16	46	.0046
17	10	.001
18	0	0
19	0	0
20	0	0
------		------
L2(22) =		

The displays below show the histogram of the frequencies of the outcomes and superimposed on the histogram, the curve with equation $y=10000(\sqrt{10\pi})^{-1}e^{\wedge}((x-10)^2/10)$

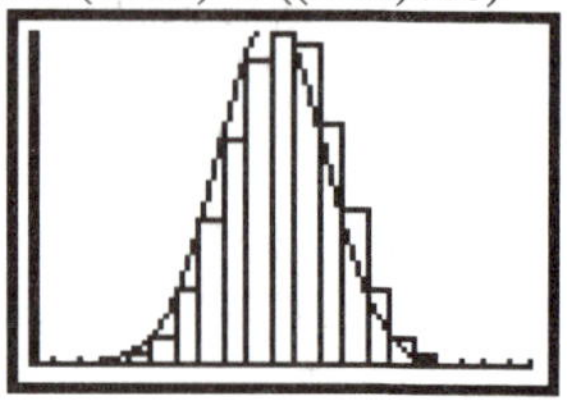

Copyright © 1998 by Brendan Kelly Publishing Inc.

Answers to Exercises & Hints for Investigations

Exploration 11

Note: In this exercise set, we revisit some problems which we tackled in earlier Explorations using simulations. Here we *calculate* (rather than estimate) theoretical probabilities, using such tools as the binomial formulas on page 47 and the binomial functions .

1. The *binomial distribution* defines the probability of achieving exactly k occurrences of an event on n independent trials of an experiment with only two possible outcomes of given probability.

The trials of an experiment are said to be *independent* if the outcome of any trial is not affected by the outcome of a previous trial. Successive samplings with replacement from a finite set represent independent trials, but samplings without replacement are not independent.

A set of trials of an experiment are said to be *Bernoulli trials* if:
•there are exactly two possible outcomes on each trial.
•each trial is independent of the others.
•the probability of each outcome is the same for each trial.

2. Using the binomial formula, we calculate the probability of exactly 10 successes on 20 trials (where the probability of success on a single trial is 0.5) to be:

$${}_{20}C_{10}(0.5)^{10}(0.5)^{10} \approx 0.1762.$$

Verification:

```
binompdf(20,.5,1
0)
        .176197052
```

3. If each card is returned to the deck after it is drawn, then we say the sampling is done "with replacement". In this case each draw is independent of the draws which preceded it, and the sampling can be regarded as a Bernoulli trial. The student's solution would be correct if the sampling is done with replacement.

If sampling is without replacement, then the probabilities on successive trials are dependent and the binomial formula does not apply.

4. a) Using the binomial formula, we calculate the probability of exactly 5 boys on 7 trials to be:

$${}_{7}C_{5}(0.5)^{5}(0.5)^{2} \approx 0.1641$$

Verification:

```
binompdf(7,.5,5)
         .1640625
```

b) We are assuming that both sexes are equally likely and that the trials are independent, i.e. the sex of the first n children does not affect the sex of the $n + 1^{th}$ child.

5. Using the binomial formula, we calculate the probability of exactly 13 "successes" on 25 trials to be:

$${}_{25}C_{13}(0.25)^{13}(0.75)^{12} \approx 0.00246.$$

Verification:

```
binompdf(25,.25,
13)
       .0024546166
```

Exploration 11 (cont'd)

6. The probability that there are *at least* 4 arrows in the inner circle is: 1 – p(no arrows) – p(one arrow) – p(2 arrows) – p(3 arrows); that is,

$$1-[0.8^{20}+20(0.2)(0.8)^{19}+{}_{20}C_{2}(0.2)^{2}(0.8)^{18}+{}_{20}C_{3}(0.2)^{3}(0.8)^{17}]$$
$$\approx 0.5885.$$

Verification:

```
1-binomcdf(20,.2
,3)
       .5885511383
```

7. a) If we draw *with replacement*, 5 birthdays from the 12 months, then the probability of exactly 2 "January" draws is the probability of 2 successes on 5 trials; i.e.

$${}_{5}C_{2}\left(\frac{1}{12}\right)^{2}\left(\frac{11}{12}\right)^{3} \approx 0.0535$$

Verification:

```
binompdf(5,1/12,
2)
       .0534899048
```

b) p(at least 2 people) = 1 – p(nobody) – p(exactly 1 person)

$$= 1-\left(\frac{11}{12}\right)^{5}-{}_{5}C_{1}\left(\frac{1}{12}\right)\left(\frac{11}{12}\right)^{4} \approx 0.0586\ .$$

Verification:

```
1-binomcdf(5,1/1
2,1)
       .0585776749
```

7. c) Reasoning as in *Worked Example* 3 page 74, we obtain:

$$1-\left(\frac{11}{12}\right)\left(\frac{10}{12}\right)\left(\frac{9}{12}\right)\left(\frac{8}{12}\right) \approx 0.618$$

d) No. The probabilities cannot be added, because the outcomes "two birthdays in January" and "two birthdays in February" are not mutually exclusive. By adding these probabilities, we count twice the cases where two of the 5 people have birthdays in January and two others have birthdays in February.

8. We compute the theoretical probability by observing that the Professor's departure times overlap the newspaper arrival times by 5 minutes. One-third of the time, the Professor will leave between 7:10 and 7:15. The newspaper arrival time will fall in that interval one time in four, so the probability that the newspaper arrival and the Professor's departure fall in the interval 7:10 to 7:15 is 1/12. But the newspaper will arrive first in only half of these cases, so the probability that the Professor will receive the newspaper before her departure is only 1/24. The probability that this will occur at least once in five trials is $1-(23/24)^{5} = 0.1916\ldots$ about 1 chance in 5. See the solution to exercise 8 of *Exploration* 10 for a simulation.

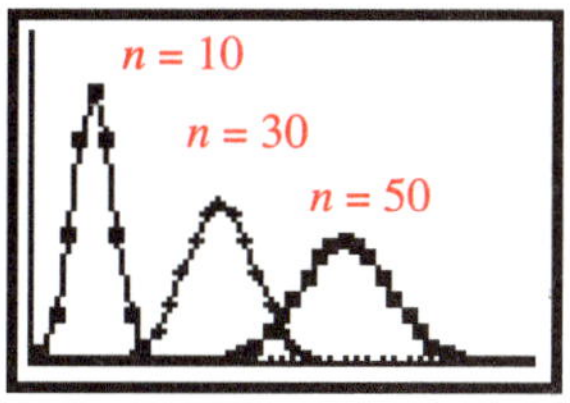

9. a) & b) We observe that as the number of coins is increased, the mean number of heads increases and the standard deviation increases.
c) $n/2$
d) The probability that exactly $n/2$ heads occurs is ${}_{n}C_{n/2}\,2^{-n}$.
Substitution from Stirling's formula yields $\sqrt{\frac{2}{\pi n}}$.

Copyright © 1998 by Brendan Kelly Publishing Inc.

Answers to Exercises & Hints for Investigations

Exploration 12

1. The equations are: $y =$

a) $\frac{0.1}{\sqrt{2\pi}}e^{-\frac{1}{2}\left(\frac{x-50}{10}\right)^2}$ b) $\frac{0.1}{\sqrt{2\pi}}e^{-\frac{1}{2}\left(\frac{x-60}{10}\right)^2}$ c) $\frac{0.1}{\sqrt{2\pi}}e^{-\frac{1}{2}\left(\frac{x-80}{10}\right)^2}$

2. The graphs of the equations in 1a), 1b) and 1c) are shown from left to right. As an alternative to entering the equations, we can use the **binompdf(** command shown on the display.

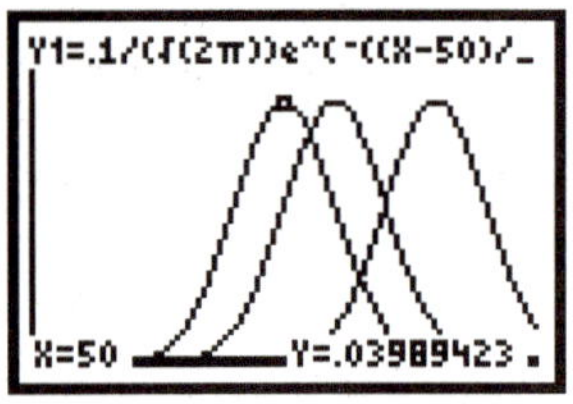

$0 \le x \le 100;\ 0 \le y \le 0.05$

```
Plot1 Plot2 Plot3
\Y1■normalpdf(X,
50,10)
\Y2■normalpdf(X,
60,10)
\Y3■normalpdf(X,
80,10)
\Y4=
```

We observe that if μ is increased by k units, the corresponding graph moves k units to the right if $k > 0$ and to the left if $k < 0$. The shape of the graph is unchanged.

3. The equations are: $y =$

a) $\frac{0.2}{\sqrt{2\pi}}e^{-\frac{1}{2}(\frac{x-50}{5})^2}$ b) $\frac{0.1}{\sqrt{2\pi}}e^{-\frac{1}{2}(\frac{x-50}{10})^2}$ c) $\frac{0.05}{\sqrt{2\pi}}e^{-\frac{1}{2}(\frac{x-50}{20})^2}$

4. The graphs of the equations in 3a), 3b) and 3c) are shown from tallest to shortest. As an alternative to entering the equations, we can use the **binompdf(** command shown on the display.

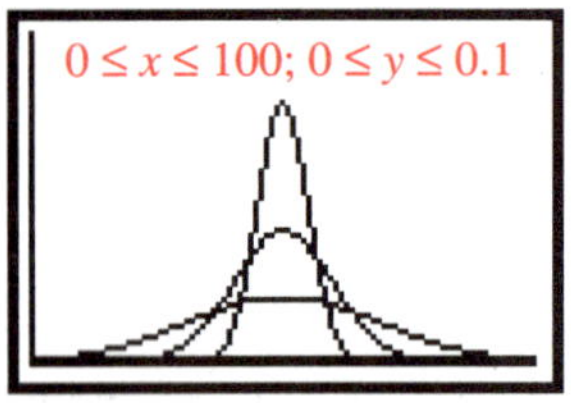
$0 \le x \le 100;\ 0 \le y \le 0.1$

```
Plot1 Plot2 Plot3
\Y1■normalpdf(X,
50,5)
\Y2■normalpdf(X,
50,10)
\Y3■normalpdf(X,
50,20)
\Y4=
```

```
Plot1 Plot2 Plot3
\Y1■.2/(√(2π))e^
(-((X-50)/5)²/2)
\Y2■.1/(√(2π))e^
(-((X-50)/10)²/2
)
\Y3■.05/(√(2π))e
^(-((X-50)/20)²/
2)
\Y4=
```

We observe that as σ is increased, by a factor of k, the height of the curve is decreased by a factor of k and the graph becomes broader, with less of its area concentrated close to the mean.

5. a) If the measurement errors are random, i.e., if it is equally likely that a measurement will be larger or smaller than the true value, then the measurements should be roughly equally distributed on both sides of the true value. Since the mean should be close to the true value, then the data should be equally distributed about the mean. That is, the data should by symmetric about the mean.

b) The values $x = \mu + \varepsilon$ and $x = \mu - \varepsilon$ are equidistant from the line $x = \mu$ and on opposite sides of it (for $\varepsilon > 0$). Substituting both values into the normal equation, we discover that they yield the same value of y.

Exploration 12 (cont'd)

6. a) and b)

$$y = \frac{1}{15\sqrt{2\pi}}e^{-\frac{1}{2}\left(\frac{x-100}{15}\right)^2}$$

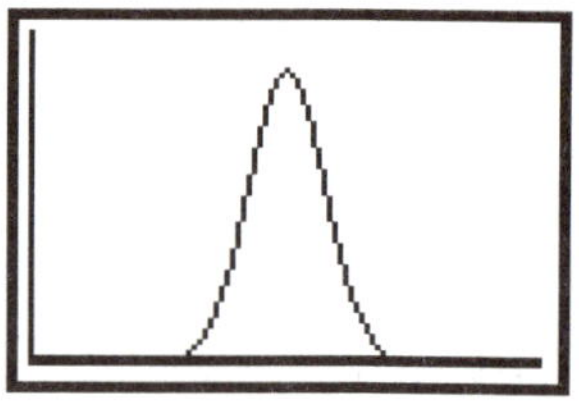
$0 \le x \le 200;\ 0 \le y \le 0.03$

The display shows how we can generate the graph using the **normalpdf(** command.

```
Plot1 Plot2 Plot3
\Y1=normalpdf(X,
100,15)
```

c) We proceed as in *Worked Example* 2 to obtain a percentage of 0.6826… between 85 and 115 and 0.2475… between 90 and 100. Alternatively, we can use the **ShadeNorm(** command as shown.

(i)

```
ShadeNorm(85,115
,100,15)
```

(ii)

```
ShadeNorm(90,100
,100,15)
```

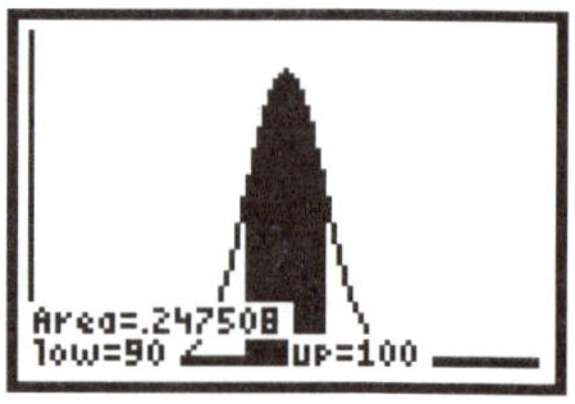

d) Trace along the curve to the points with half the maximum height, or intersect the graph with the line $y = 0.0133$ (half the max height). We find a frequency of N/2 occurs at marks of 82 and 118.

7. c) Use the range settings: $120 \le x \le 200;\ 0 \le y \le 0.08$
We evaluate the area between 120 and 160 to obtain about 34%

d) As in *Worked Example* 3, we use the inverse norm command to find that the 95th percentile is 172.5.

```
invNorm(.95,162.
5,6.1)
         172.5336071
```

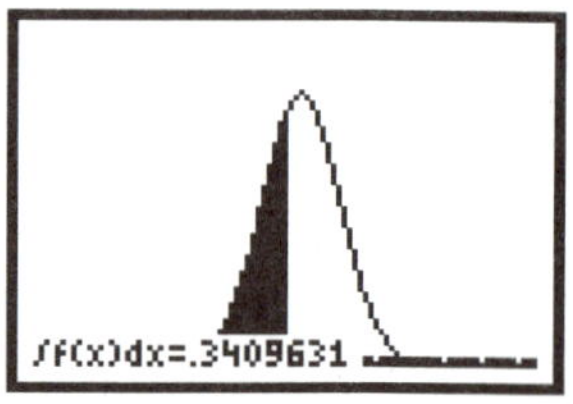

8. a & b)
Using our estimates of μ and σ, we obtained the equation of the probability function for the SAT scores shown in the display below.
When we graphed this equation and the corresponding xyLine with window variables as shown, we obtained the display shown here.

```
WINDOW
 Xmin=200
 Xmax=800
 Xscl=50
 Ymin=0
 Ymax=.004
 Yscl=1
 Xres=1
```

```
Plot1 Plot2 Plot3
\Y1■(125√(2π))⁻¹e
^(-((X-504)/125)
²/2)
```

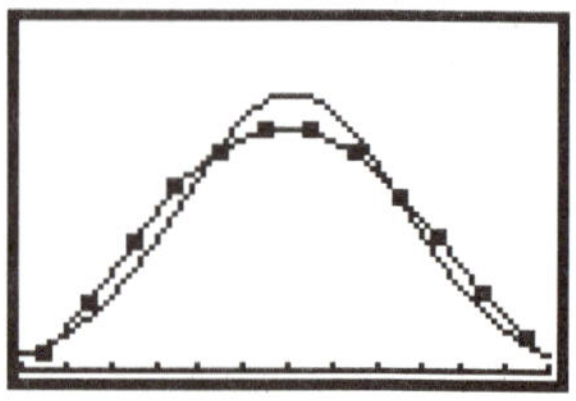

d) The area under the normal curve between 400 and 600 is about 0.576… which is close to 52.7%.

e) To find the 95th percentile, we use the **invNorm(** command. We find the 95th percentile is approximately 710.

Copyright © 1998 by Brendan Kelly Publishing Inc.

Answers to Exercises & Hints for Investigations

Exploration 13

1.

$$t_5(x)=\frac{C_5}{\left(1+\frac{x^2}{5}\right)^3},\quad t_{20}(x)=\frac{C_{20}}{\left(1+\frac{x^2}{20}\right)^{21/2}},\quad t_{50}(x)=\frac{C_{50}}{\left(1+\frac{x^2}{50}\right)^{51/2}}$$

2. a)

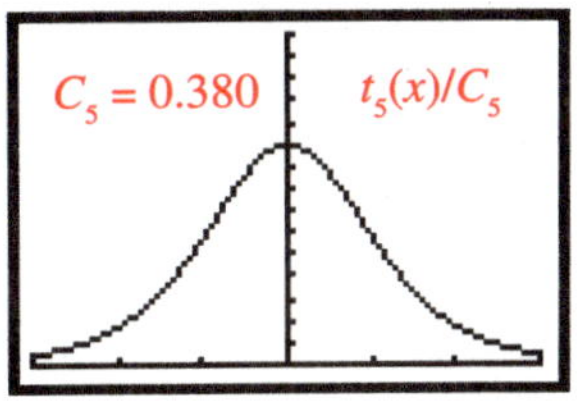

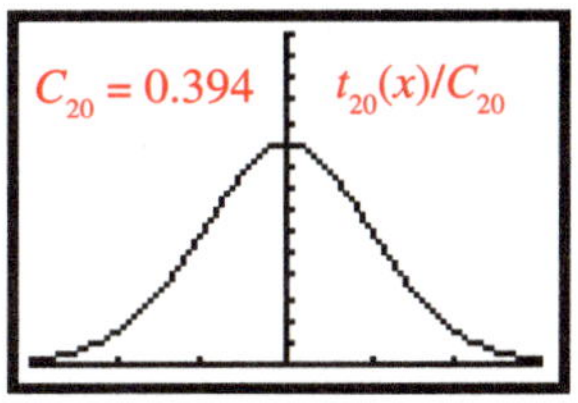

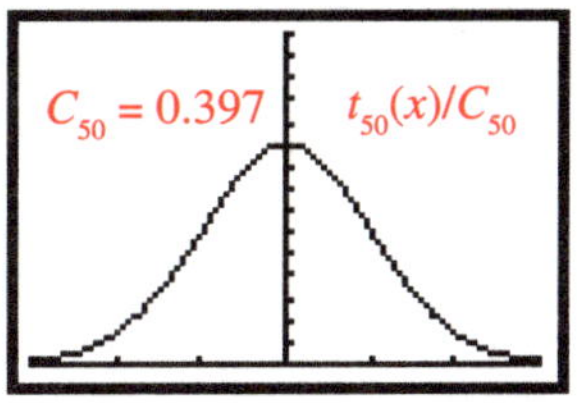

b) The display below shows the three graphs together. As n increases, C_n approaches the value 0.399 which is the constant in the normal density function. In fact, as $n \to \infty$, $t_n(x)$ approaches the normal distribution. For small values of n, the graph of $t_n(x)$ is slightly wider and not quite as tall as the graph of the normal distribution.

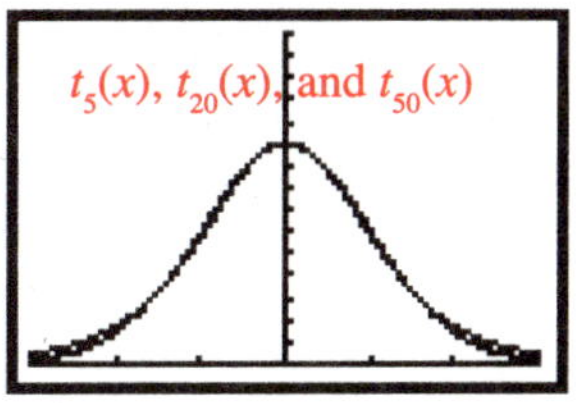

3. Since the t distribution is symmetric about the y-axis, half the area under the graph occurs in the second quadrant, so Prob $t_n(x) < 0$ is 0.5 for *any* value of n.

4. a) Substitution of $-x$ into the expression for $t_n(x)$ yields $t_n(x)$.

b) $t_n(x)$ is symmetric about the y-axis.

c) Since $t_n(x_0) = t_n(-x_0)$, the values $-x_0$ and x_0 are equally likely. Therefore, the mean value of $t_n(x)$ is 0.

5. The graphs in the displays are drawn in the window: $-3 \le x \le 3$; $0 \le y \le 0.5$ The command used for each graph is shown above it.

a)
```
Shade_t(-1,1,5)
```

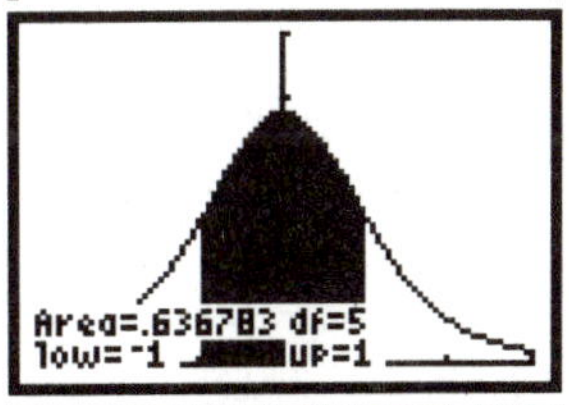

b)
```
Shade_t(-.5,1,8)
```

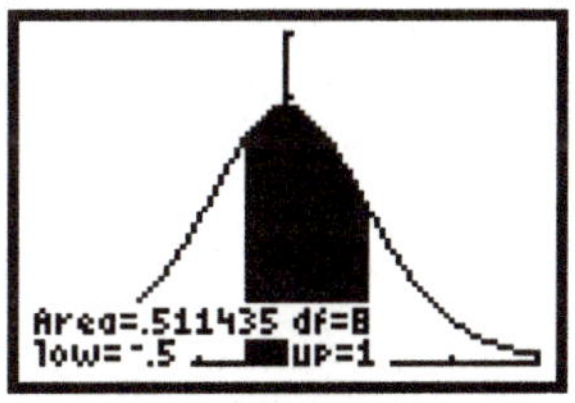

c)
```
Shade_t(-1E99,1,
10)
```

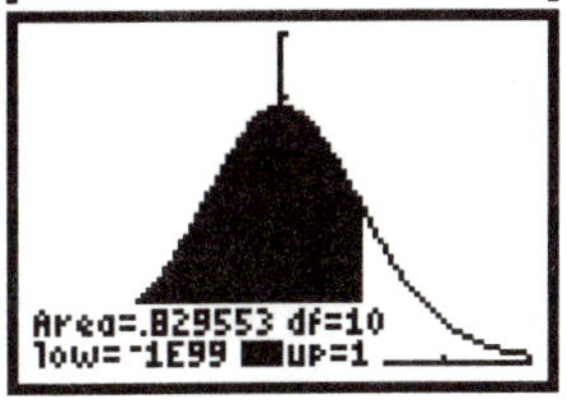

d)
```
Shade_t(0.8,1E99
,20)
```

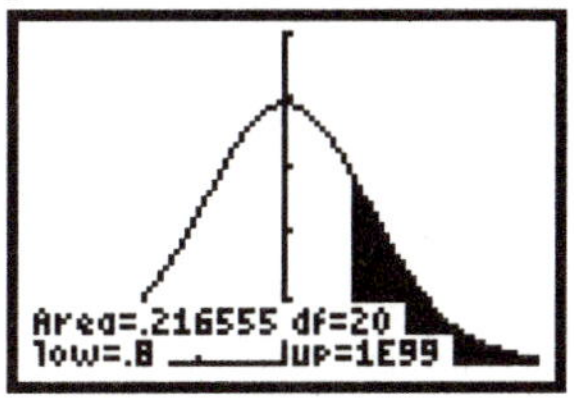

Exploration 13 (cont'd)

6. The probability is about 0.949 as shown in the display.

```
tcdf(-1E99,1.8,1
0)
        .9489738803
```

7. a) The 50th percentile is the value of x for which tcdf(x) = 0.5. Since the t distribution is symmetric about the y-axis, exactly half its area is to the left of the y-axis, so the 50th percentile is 0.

b) The trial-and-error method will yield different answers depending upon how close your estimates are to the true value. The actual 60th and 95th percentiles are respectively 0.2586 and 1.7709. The displays below show some approximations using the trial-and-error method. See investigation 9 for a method of calculating precise percentiles.

```
tcdf(-1E99,.26,1
3)
         .600531911
```

```
tcdf(-1E99,1.8,1
3)
        .9524505344
```

8. a) $C_1 = 1/\pi$ $\quad C_3 = 2/(\pi\sqrt{3})$ $\quad C_{30} = 29!\sqrt{30}\,/\,(2^{30}\,15!14!)$

b) $C_1 \approx 0.318$ $\quad C_3 \approx 0.368$ $\quad C_{30} \approx 0.396$

c) $$t_3(x) = \frac{2}{\pi\sqrt{3}\left(1+\frac{x^2}{3}\right)^2}$$

When we graph $t_3(x)$ we obtain the graph shown in the display. We obtain the same graph when we plot **tpdf(X, 3)**.

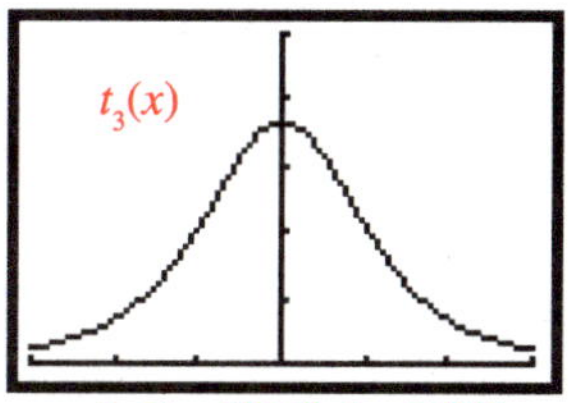

$-3 \le x \le 3;\ 0 \le y \le 0.5$

For Calculus Buffs

- $\displaystyle\int_{-\infty}^{\infty}\frac{dx}{1+x^2} = \tan^{-1}x\Big|_{-\infty}^{\infty} = \pi$, so $C_1 = 1/\pi$.

- After the appropriate continuity conditions have been established, our proof that the t-distribution approaches the normal distribution follows from the definition of e.

$$\lim_{n\to\infty} t_n(x) = C\lim_{n\to\infty}\int\left(1+\frac{x^2}{n}\right)^{-\left(\frac{n+1}{2}\right)}dx$$

$$= C\int\lim_{n\to\infty}\left(1+\frac{x^2}{n}\right)^{-\left(\frac{n+1}{2}\right)}dx$$

$$= C\int\left[\lim_{n\to\infty}\left(1+\frac{x^2}{n}\right)^n\right]^{-\left(\frac{1}{2}+\frac{1}{2n}\right)}dx$$

$$= C\int e^{-x^2/2}dx$$

- $$\int\frac{1\,dx}{\left(1+\frac{x^2}{3}\right)^2} = \frac{3x}{2(x^2+3)} + \frac{\sqrt{3}\tan^{-1}(x/\sqrt{3})}{2} + C$$

Copyright © 1998 by Brendan Kelly Publishing Inc.

Answers to Exercises & Hints for Investigations

Exploration 14

1. $\chi^2_7(x) = K_7 x^{5/2} e^{-x/2}$, $\chi^2_9(x) = K_9 x^{7/2} e^{-x/2}$, $\chi^2_{10}(x) = K_{10} x^4 e^{-x/2}$

2. a) The graphs are shown in the display. Following the procedures of *Worked Example* 1, we obtain these values of K: $K_7 = 0.0265$, $K_9 = 0.00380$, $K_{10} = 0.00130$.

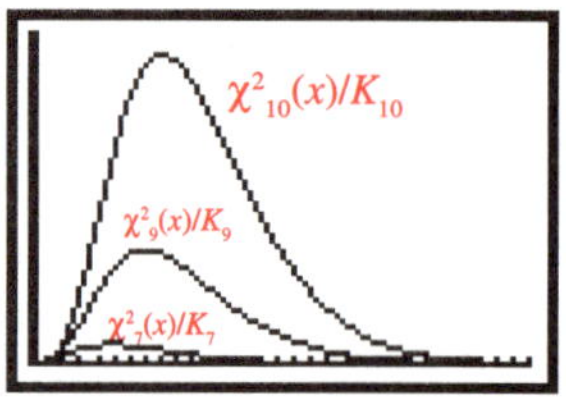

b) When we insert these constants and graph the χ^2 functions, in the range $0 \le x \le 30$; $0 \le y \le 0.15$, we obtain the graphs shown in the display. As the number of degrees of freedom increases, the curve becomes more spread out and its peak is reduced in height. This shows that large values are more probable for chi-square distributions with a large number of degrees of freedom.

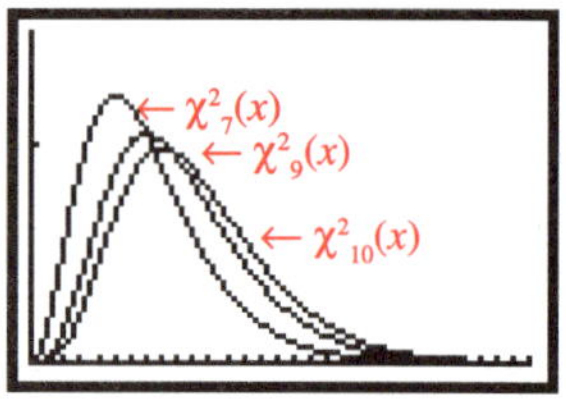

3. When we graph $\chi^2_8(x)$ and evaluate the maximum, we obtain the maximum probability density of 0.112 when $x = 6$, as shown in the display.

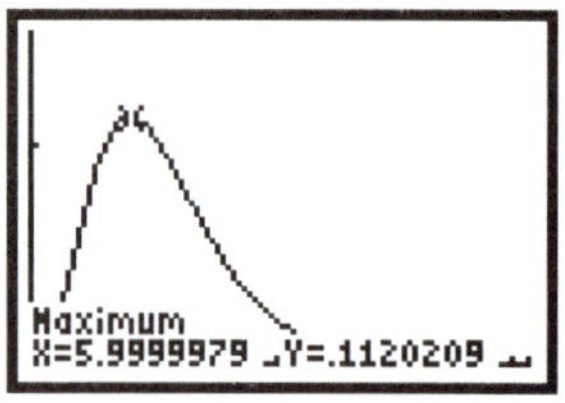

4. a) The trial-and-error method will yield different answers depending upon how close your estimates are to the true value. The actual 50th percentiles are respectively 12.34, 14.338, 24.336, and 29.336. The displays below show some approximations using the trial-and-error method. See investigation 9 for a method of calculating precise percentiles.

```
X²cdf(0,12.3,13)
        .4967893432
```

```
X²cdf(0,14.3,15)
        .4970910819
```

```
X²cdf(0,24.3,25)
        .4979020042
```

```
X²cdf(0,29.3,30)
        .4981190293
```

b) The table shows that the median of each distribution seems to be 0.7 less than the number of degrees of freedom.

MED	DOF
12.3	13
14.3	15
24.3	25
29.3	30

c) The relationship in b) suggests that the median of the chi-square with 40 degrees of freedom is about 39.3. The display verifies this relationship yields a good estimate.

```
X²cdf(0,39.3,40)
        .498407512
```

5. The displays show the areas in the indicated window.

a)

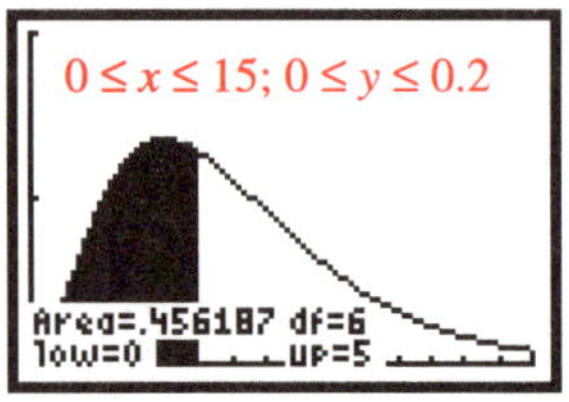

b)

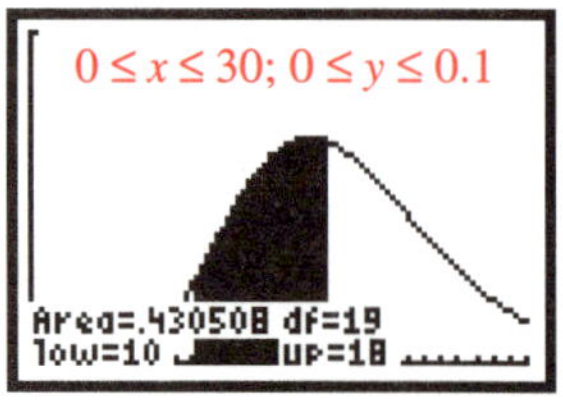

Exploration 14 (cont'd)

5. c)

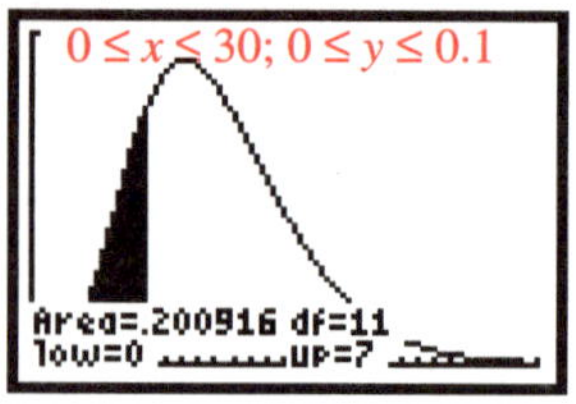

d)

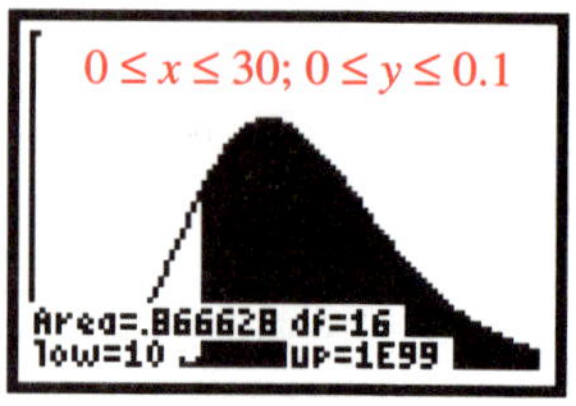

6. The display shows that the probability the $\chi^2_{10}(x)$ assumes a value less than 9 is 0.467…

```
X²cdf(0,9,10)
        .4678964231
```

7. The actual 60th and 95th percentiles are 13.635 and 22.362.

8. a) $K_1 = \dfrac{1}{\sqrt{2\pi}}$ $\quad K_4 = \dfrac{1}{4}$ $\quad K_5 = \dfrac{1}{3\sqrt{2\pi}}$

b) $K_1 = 0.3989\ldots$ $\quad K_4 = 0.25$ $\quad K_5 = 0.1329\ldots$

c) The equation for $\chi^2_5(x)$ is $\chi^2_5(x) = 0.1329x^{3/2}e^{-x/2}$. We obtain the same graph with the command

χ²pdf(X,5).

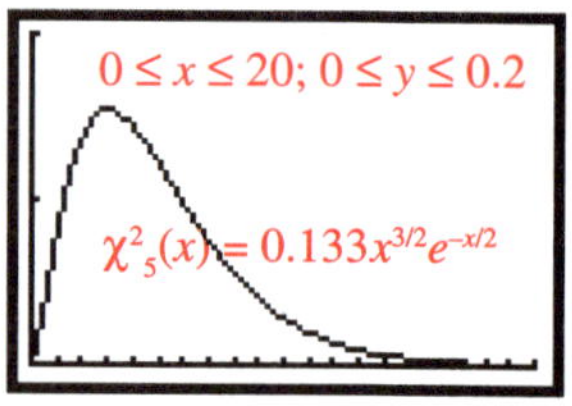

Exploration 15

1. The following functions are defined for $x > 0$, and 0 elsewhere.

a) $f_{4,6}(x) = \dfrac{k}{\left(1+\dfrac{2}{3}x\right)^5} x$ $\qquad$ b) $f_{5,6}(x) = \dfrac{k}{\left(1+\dfrac{5}{6}x\right)^{11/2}} x^{3/2}$

c) $f_{5,7}(x) = \dfrac{k}{\left(1+\dfrac{5}{7}x\right)^6} x^{3/2}$

2. a) $k_{4,6} = 5.333\ldots$ $\qquad k_{5,6} = 12.481\ldots$ $\qquad k_{5,7} = 11.712$

b) If we graph all three functions in the window: $0 \le x \le 5; 0 \le y \le 1$, the graphs are hardly distinguishable. To see the effect of increasing n_1 while holding n_2 constant, enter $\{3, 5, 15, 40\} \rightarrow L_1$. Then define a family of F distributions having this set of values of n_1 and fixed $n_2 = 6$ as shown in the display on the left. We then hold n_1 fixed at 5 and let n_2 run through the values in the list L_1.

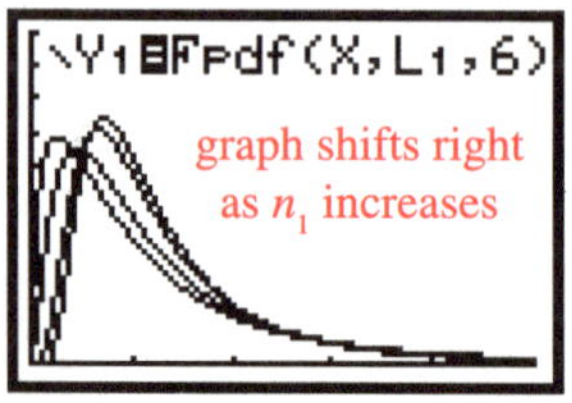

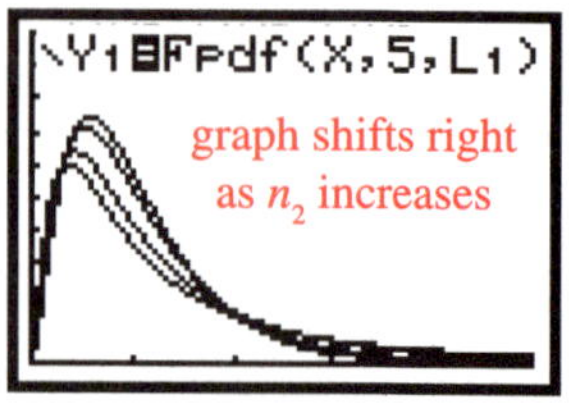

Copyright © 1998 by Brendan Kelly Publishing Inc.

Answers to Exercises & Hints for Investigations

Exploration 15 (cont'd)

3. When we graph $F_{8,10}$ and evaluate the maximum, we obtain the maximum probability density of 0.728 when $x = 0.625$, as shown in the display.

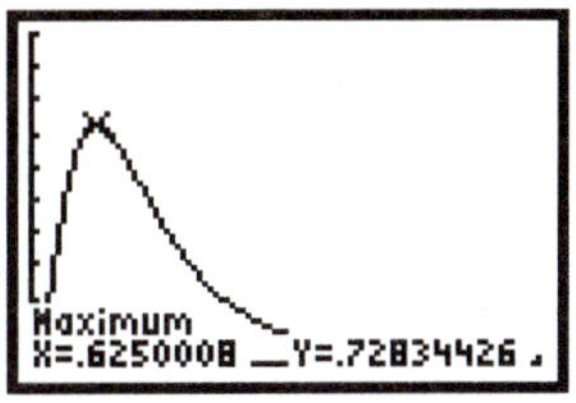

4. a)

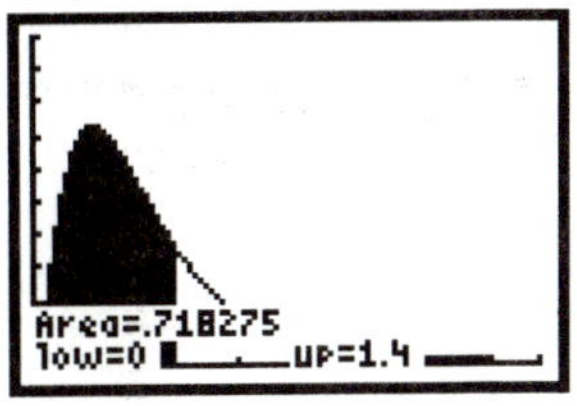

b)

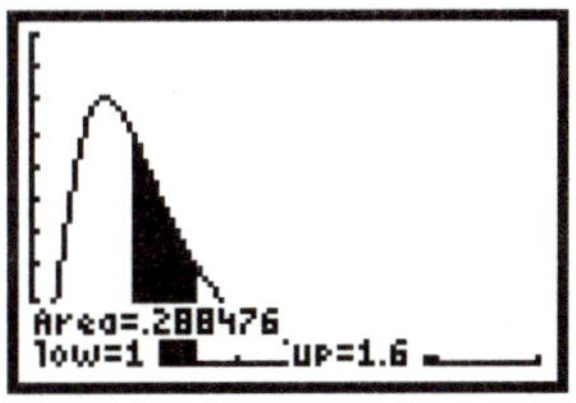

c)

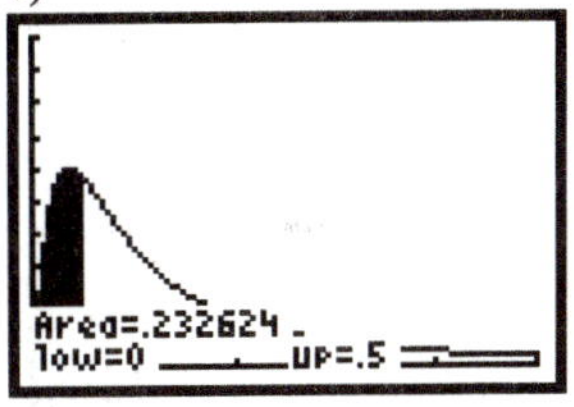

d)

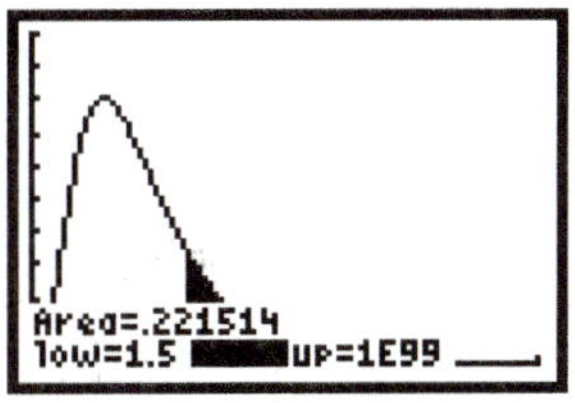

5. The display shows that the probability the $F_{10,13}(x)$ assumes a value less than 2 is 0.879…

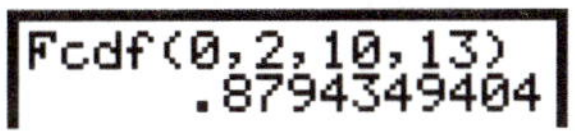

6. a) F(1) – F(0) b) F(1.8) – F(0) c) F(1.8) – F(1)
d) F(∞) – F(0) e) 1 – F(1.8)
f) If $F(x) = 0$ for $x \le 0$, then F(0) = 0 and part d) is equal to 1.

7. a) We define the cumulative distribution function of $F_{5,12}$ as shown in the display.

b) The graph of Y_1 as defined in part a) is shown in the display. When this graph is evaluated at $x = 1.8$, it yields $Y_1 = 0.8229$ as shown. That is, F(1.8) = 0.8229. Similarly we find F(1) = 0.5238… and so, F(1.8) – F(1) = 0.2991. This verifies the area displayed in exercise 6.

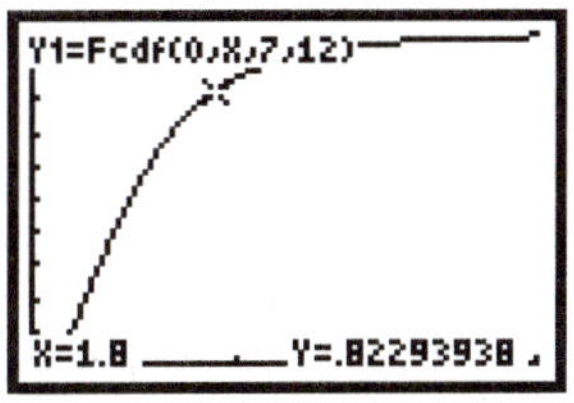

c) The display shows further verification.

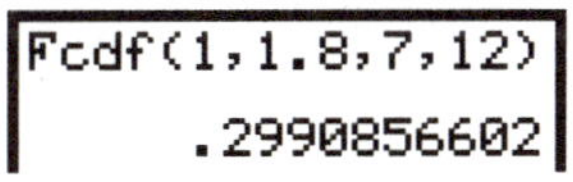

8. a) & b) The displays show that the 90th and 95th percentiles are respectively $\alpha = 1.9877\ldots$ and $\beta = 2.4281\ldots$
Furthermore, $F(\beta) - F(\alpha) = 0.05$. (as we might expect)

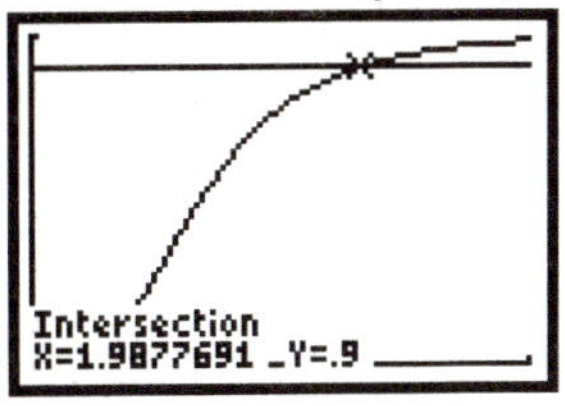

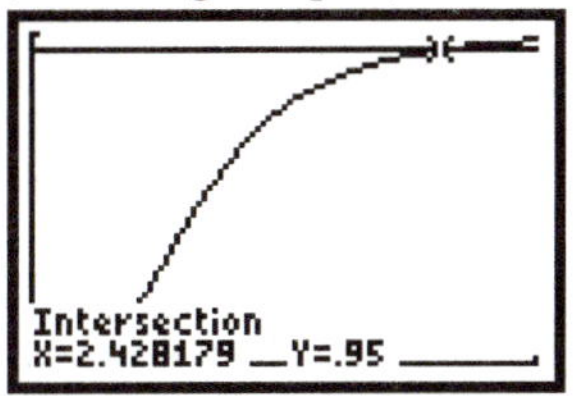

Exploration 16

1. The theorem on page 63 indicates that we must include a "change of variable" factor $\left|\frac{dx}{dy}\right|$ as shown in *Worked Example* 3.

2. a) $F_X(-\sqrt{x})$ b) $1 - F_X(\sqrt{x})$ c) $F_X(\sqrt{x}) - F_X(-\sqrt{x})$
d) Since $f_X(x)$ is symmetric about the y-axis, the mapping $x \to -x$ maps the left tail of the graph onto the right tail, so they are congruent and their areas are equal; i.e. $F_X(-\sqrt{x}) = 1 - F_X(\sqrt{x})$.
e) $F_Y(x) = \text{Prob}(-\sqrt{x} \le x \le \sqrt{x}) = F_X(\sqrt{x}) - F_X(-\sqrt{x})$
$= F_X(\sqrt{x}) - [1 - F_X(\sqrt{x})] = 2F_X(\sqrt{x}) - 1$.

3. In *Worked Example* 2 (and in exercise 2), we proved that if X is a random variable whose probability density function is symmetric about the y-axis, then $F_Y(x) = 2F_X(\sqrt{x}) - 1$ when $Y = X^2$. Using this relationship, we define the cumulative distribution of Y, shown as Y_1 in the display. We define the cumulative distribution $F_{1,8}(x)$ as Y_2 in the display. When we graph these two distributions, we discover that their graphs are coincident; that is, the distribution of the square of a t distribution with 8 degrees of freedom is the F distribution with 1 and 8 degrees of freedom.

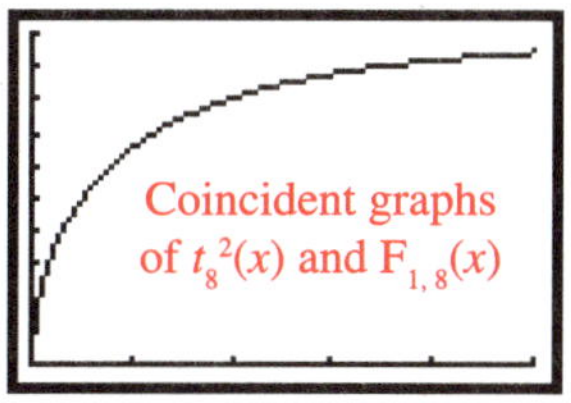

4. a) Following the steps in *Worked Example* 1 we obtain:
$$f_Y(x) = \frac{1}{5} f_X(x/5).$$
b) Substitute $x = 2.5$ into equation in a) to obtain $f_Y(2.5) = 0.008$.

6. For any $x > 0$, either $X \le x$ or $X > x$. That is,
Prob $(X \le x)$ + Prob $(X > x) = 1$. But $X > x$ if and only if $\frac{1}{X} < \frac{1}{x}$.
Therefore, Prob $(X \le x)$ + Prob $\left(\frac{1}{X} < \frac{1}{x}\right) = 1$.

7. a) & b) The graphs intersect at $x = 1$. For $x > 1$, each graph seems to be a reflection of the other in the line $y = 1$.

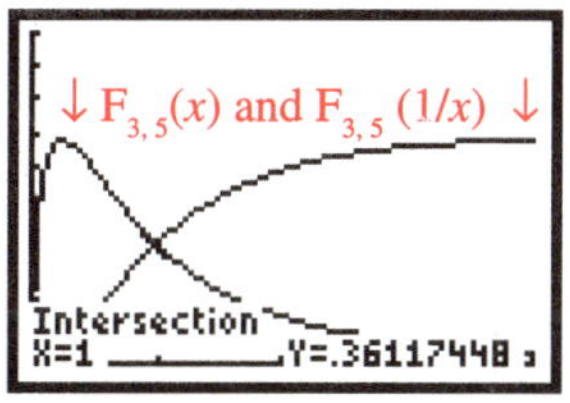

8. a) The 95th percentile is 5.40945 The 5th percentile is 0.18486

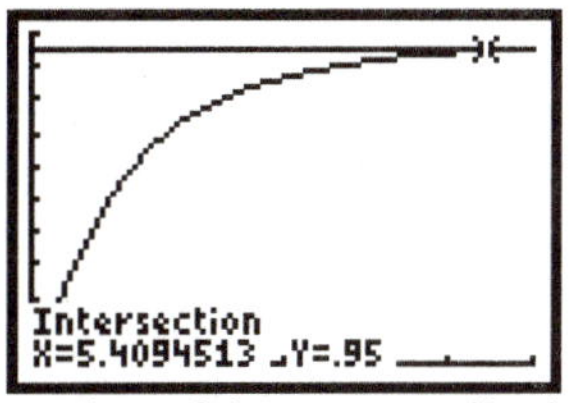

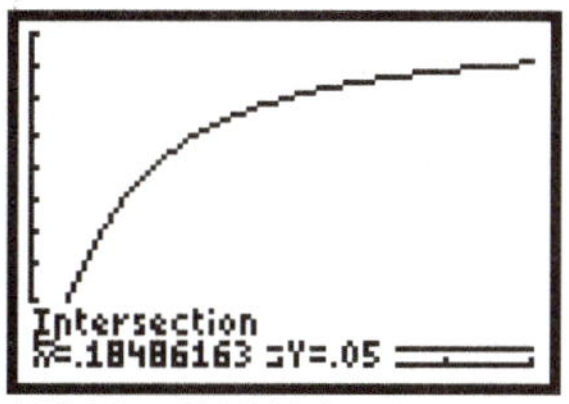

The product of these percentiles is $5.40945 \times 0.18486 = 1$

b) The 90% confidence interval is: $0.11094 \le x \le 5.40945$.

Copyright © 1998 by Brendan Kelly Publishing Inc.

Answers to Exercises & Hints for Investigations

Exploration 17

1. a) The display below left shows that the probability is about 0.965
b) When we attempt to evaluate **binomcdf(** with n = 10,000, we get the error message shown below right.

The **binomcdf(** reaches overload when n is too large.

2. The display shows that the probability of more than 275 heads is about 0.0112.

3. We approximate the binomial distribution with mean 250 and standard deviation $\sqrt{125} \approx 11.18$, with the normal distribution having mean 250 and standard deviation 11.18. Then the probability sought is the area $1 - F_H(275)$. The displays show the area and indicate that the probability of more than 275 heads is about 0.0126. This is a little larger than probability 0.0112 found in Exercise 2.

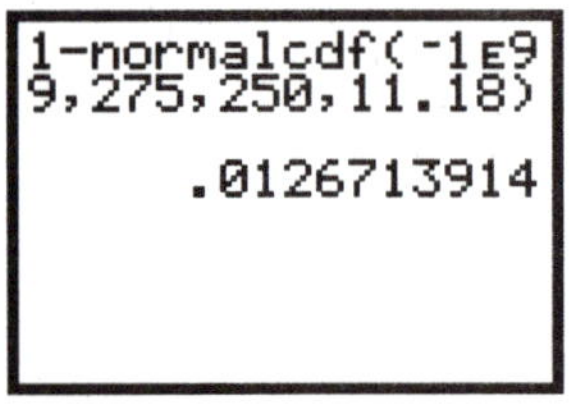

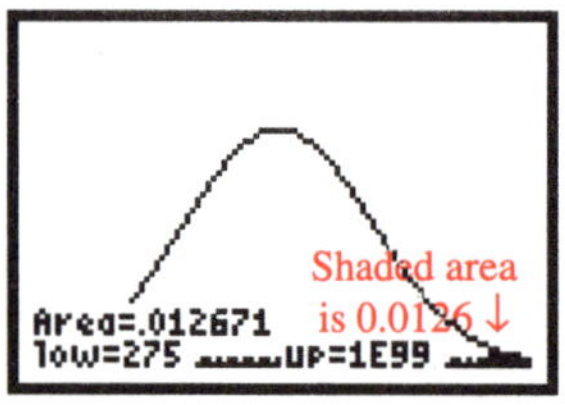

4. The Central Limit Theorem tells us that it is not necessary that the random variables $X_1, \ldots X_n$ be normally distributed. As long as n is large (greater than 25), the distribution of the mean of these variables will be close to a normal distribution, with the approximation improving for larger values of n.

5. This exercise requires that we use the binomial cumulative distribution function to find the interval that contains the middle 90% of all the area under the graph of the binomial function. The interval must exclude the right and left "tails" of the graph, which each contain 5% of the total area. These tails are the 5th and 95th percentiles. To find the 95th percentile, we graph the cumulative distribution function of the binomial distribution with mean 0.5 and standard deviation $\sqrt{25}$ or 5, and intersect it with the line $y = 0.95$ as in Investigation 9 p. 56. The display shows that the probability that there are more than 58 heads is 0.05. By the symmetry of the binomial distribution, it follows that the probability that there are fewer than 42 heads is also 0.05. That is, $h = 8$, and the required interval is $42 \le H \le 58$.

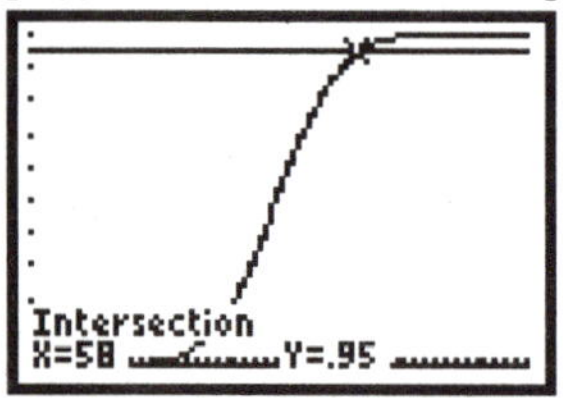

Exploration 17 (cont'd)

6. a) The null hypothesis is: *The mean lifetime of the bulbs in the entire population is 8000 hours.* We access the Z-test as in *Worked Example* 2 and then enter the values as shown in the display below left. We obtain the display shown which indicates that if the true mean is 8000 hours, then the probability that the observed mean is 48 or more hours *different from* the true mean is only 0.01318. Hence we would *not* reject the null hypothesis at the 0.01 or 1% level of significance, but we would reject it at the 5% level of significance.

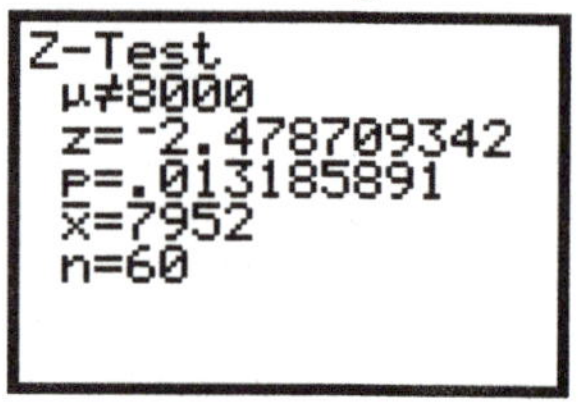

b) In this case the alternative hypothesis is "The true mean is less than 8000." This is a *one-tail test*; i.e., we are interested only in the probability that a sample of 60 bulbs will turn up a mean of 7952 or *less*. We proceed as in Exercise 6, but select the option: $< \mu_0$ opposite μ. The input and output are shown in the displays below.

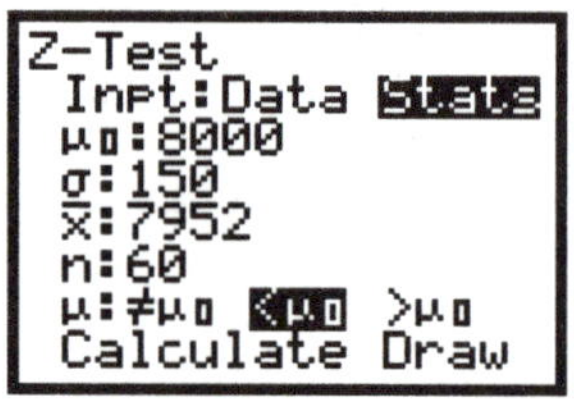

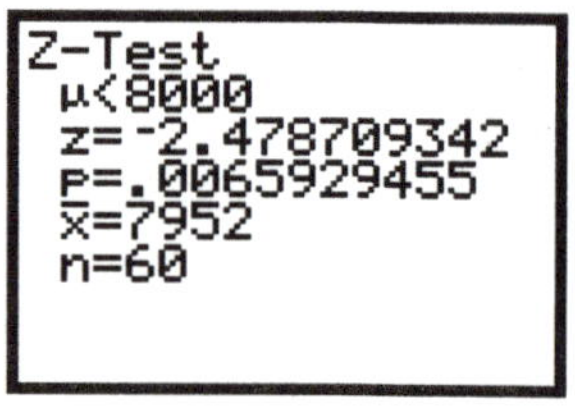

The probability that the null hypothesis is true while the measured mean is 7952 or less is 0.00659, which is smaller than the probability in part a). Thus we have enough evidence to reject the null hypothesis ($p < 0.01$), so the lifetime is very likely less than advertised.

7. a) $(1 + 2 + 3 + 4 + 5 + 6)/6 = 3.5$
b) $\sigma^2 = (1-3.5)^2 + (2-3.5)^2 + \ldots + (6-3.5)^2 = 17.5$ so $\sigma = \sqrt{17.5}$
c) If the die is fair, the mean should be close to 3.5. We use the Z-test menu as in *Worked Example* 2. The results are shown.

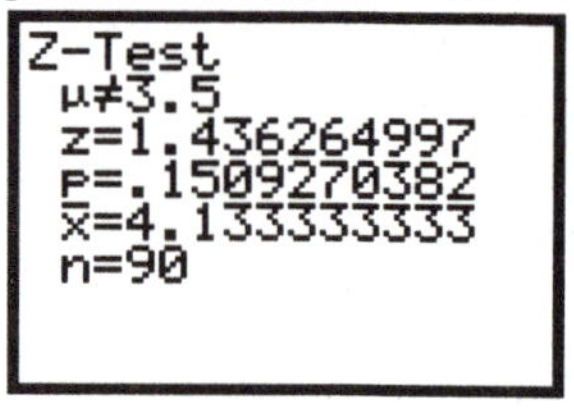

We see that there is a probability of 0.15 that this mean could occur by chance. This is greater than the 0.01 probability required to reject the null hypothesis. Therefore, we do not reject the null hypothesis.

8. b) On substitution for s^2 the left side yields

$$(n-1)s^2/\sigma^2 = \frac{\sum_{i=1}^{n}(x_i - \bar{x})^2}{\sigma^2} = \sum_{i=1}^{n}\left(\frac{x_i - \bar{x}}{\sigma}\right)^2 \quad \text{which is } \chi^2_{n-1}$$

d) We use the value of s (not σ) given on page 19, $S_x = 1.3526$, and the display indicates that the probability that the difference comes from chance variation is 0.0521, so this is not strong enough for rejection at the 5% level of significance.
f) The Z-test yields $p = 0.0394$, suggesting we reject the hypothesis; however, the Z-test is not applicable because σ is unknown.

Copyright © 1998 by Brendan Kelly Publishing Inc.

Answers to Exercises & Hints for Investigations

Exploration 18

1. Since a 1% level of significance would require a test statistic farther from zero, it is less likely that the statistic will assume assume this large a value and cause the hypothesis to be rejected.

2. a) Application of a 2-SampleZTest indicates that there is a 0.00177 probability that the difference in state and national scores is a result of chance error. Therefore the difference *is* statistically significant at the 1% level of significance.

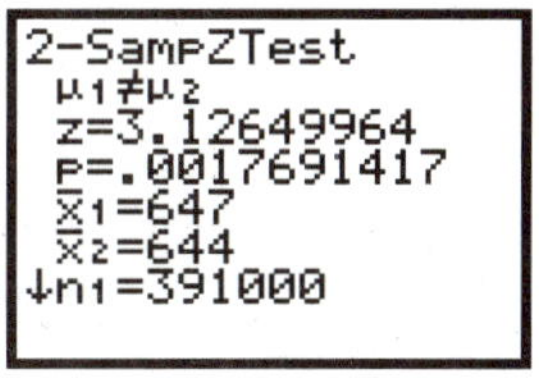

b) Whether this difference is significant in a non-statistical sense is open to opinion.

3. When the standard deviations of the populations from which the samples are drawn are known, we use the 2-Sample Z-test, but when the standard deviations are unknown, we must estimate the standard deviations and use the 2-Sample T-test.

4. a) When we substitute $n_1 = n_2 = n$ into the expression for s_p we get $s_p = \sqrt{\dfrac{s_1^{\,2} + s_2^{\,2}}{2}}$, so $s_p = \sqrt{(0.32^2 + 0.33^2)/2} \approx 0.325$

b) When we substitute s_p for σ, and apply the 2-Sample Z-test, we obtain the display showing P = 9.13×10^{-19}. The displays below compare the results of a 2-Sample T-test with a 2-Sample Z-test. A comparison reveals that the normal distribution underestimates the risk of error when the sample sizes are small because it does not allow for the error in the estimate of σ.

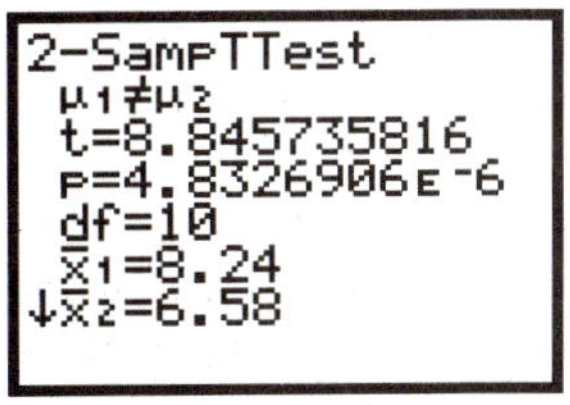

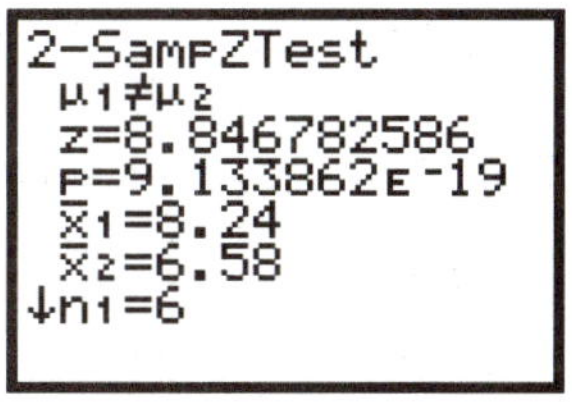

5. We enter the 30 lifetimes for Brand A into L_1 on the STAT EDIT menu, and the 30 lifetimes for Brand B into L_2.
Then we access the 2-Sample T-Test and select **Data** and ≠ μ2, and **Pooled: Yes**.

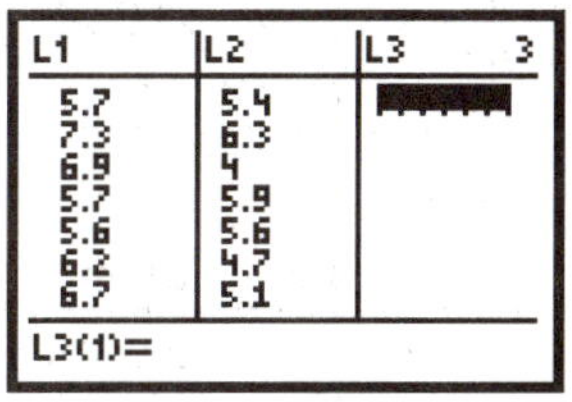

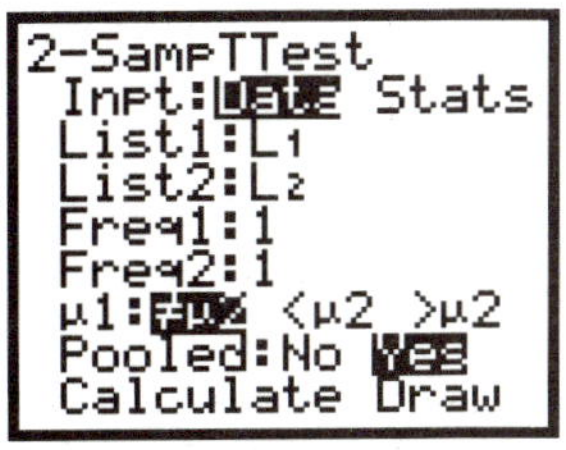

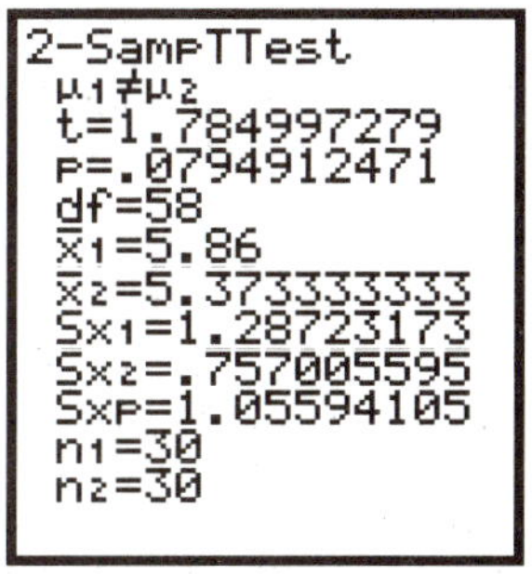

We obtain the display on the right showing that the probability P ≈ 0.079, which is greater than 0.05, so we do not have a small enough P-value to reject the null hypothesis at the 5% level of significance.

Exploration 18 (cont'd)

6. The displays show the input and results of a 2-Sample T-test.

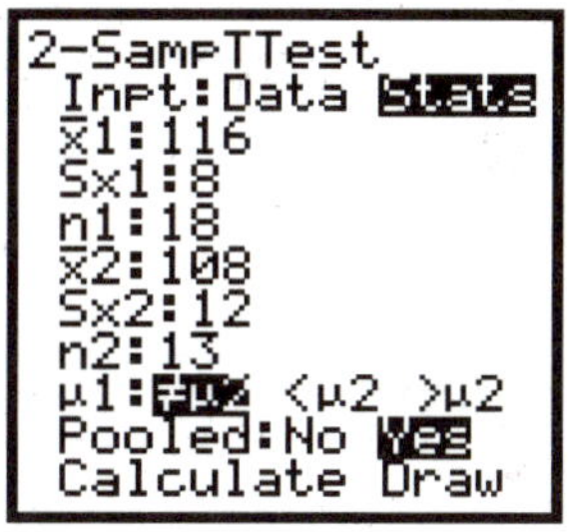

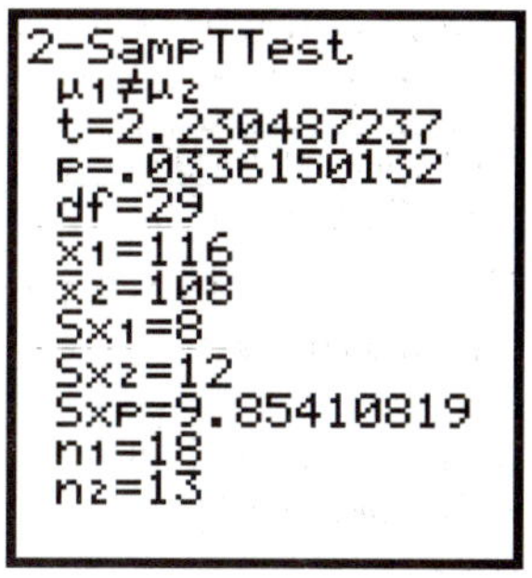

We see that P = 0.0336…or 3.36%. Therefore, we reject the hypothesis at the 5% level of significance, but cannot reject it at the 1% level of significance.

7. The displays show the input and results of a 2-Sample T-test.

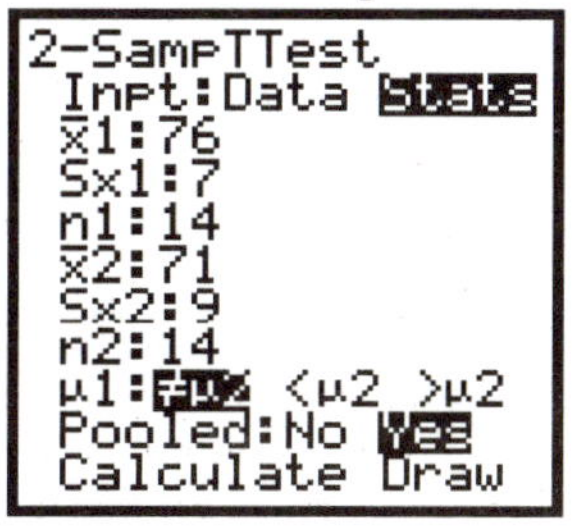

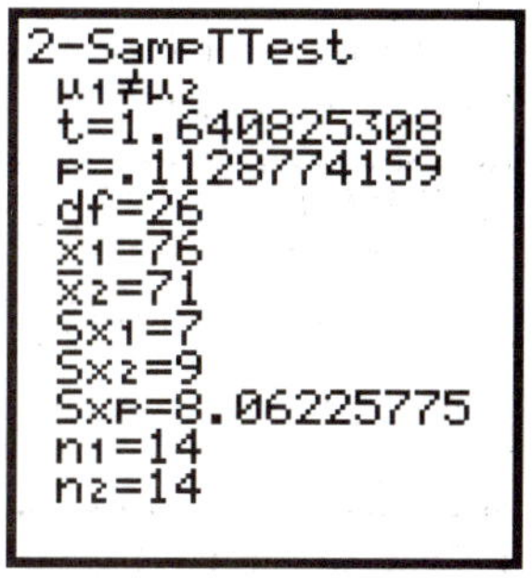

We observe that P ≈ 0.113. That is, P > 0.1, so we cannot reject at the 10% level of significance the hypothesis that the means of the two populations are equal. That is, the teacher was unable to conclude, at the 10% level of significance, that the observed difference was a real difference.

8. a) We can think of a survey on whether or not a person has access to the Internet as a sampling in which a "yes" response is a "success". We can regard X_1 (and X_2) as a sample variable that represents the number of successes in a sample of size n_1 (and n_2) from the population of males (females). The null hypothesis asserts that both populations have the same proportion p of people with access to the Internet. Since the sampling is a set of Bernoulli trials (see p. 47), X_1 and X_2 have binomial distributions with means n_1p and n_2p respectively, and corresponding standard deviations $\sqrt{n_1p(1-p)}$ and $\sqrt{n_2p(1-p)}$. And so, X_1/n_1 and X_2/n_2 have mean p and standard deviations $\sqrt{p(1-p)/n_1}$ and $\sqrt{p(1-p)/n_2}$.
For sufficiently large n_1 and n_2 we can approximate these binomial distributions with normal distributions having these means and standard deviations.
For large n_1 and n_2, $\dfrac{X_1}{n_1} - \dfrac{X_2}{n_2}$ is approximately normal with mean 0 and standard deviation, $\sqrt{p(1-p)/n_1 + p(1-p)/n_2}$.

b) It follows from part a) that the statistic below is approximately normally distributed with mean 0 and standard deviation 1, so we can use it in a Z-test of the proportions.

$$\frac{\dfrac{X_1}{n_1} - \dfrac{X_2}{n_2}}{\sqrt{p(1-p)/n_1 + p(1-p)/n_2}}$$

Copyright © 1998 by Brendan Kelly Publishing Inc.

Answers to Exercises & Hints for Investigations

Exploration 19

3. When we compute $\chi^2 = \sum_{i=1}^{6} \frac{(f_i - e_i)^2}{e_i}$, we obtain

$$\frac{(42-35)^2}{35} + \frac{(31-35)^2}{35} + \frac{(45-35)^2}{35} + \frac{(23-35)^2}{35} + \frac{(29-35)^2}{35} + \frac{(40-35)^2}{35}$$

$= 10.571\ldots$

The probability that χ^2_5 takes a value this large or larger by chance is shown on the display to be about 0.061, so we cannot reject the null hypothesis at the 1% level of significance.

```
X²cdf(10.571,E99
,5)
          .0605814213
```

Alternatively, we use the χ^2 test on the TI-83 (see *Investigation* 9) and define the matrix [A] as shown in the display.

```
MATRIX[A]  2 ×6
[ 42    31    45    23    29    40   ]
[ 1E12  1E12  1E12  1E12  1E12  1E12 ]
```

When we run the χ^2 test, we obtain the results in the display verifying the answer given above.

```
X²-Test
 X²=10.57142857
 P=.0605715018
 df=5
```

4. Using the binomial distribution (see *Exploration* 11) we calculate the expected frequencies of gender distributions to be:

4 girls 0 boys	3 girls 1 boy	2 girls 2 boys	1 girl 3 boys	0 girls 4 boys
16	64	96	64	16

The statistic $\chi^2 = \sum_{i=1}^{5} \frac{(f_i - e_i)^2}{e_i}$

$$= \frac{(11-16)^2}{16} + \frac{(40-64)^2}{64} + \frac{(94-96)^2}{96} + \frac{(76-64)^2}{64} + \frac{(35-16)^2}{16}$$

$= 35.4166$

has a χ^2 distribution with 4 degrees of freedom. The display shows that χ^2_4 takes a value of 35.4166 or larger, with a probability of about 3.814×10^{-7}, so we must reject the hypothesis that both genders are equally likely.

Alternatively, we use the χ^2 test on the TI-83 (see *Investigation* 9) and define the matrix [A] as shown in the display.

```
MATRIX[A]  2 ×5
[ 11     40     94     76     35    ]
[ 1.6E13 6.4E13 9.6E13 6.4E13 1.6E13 ]
```

The display verifies the result obtained above.

```
X²-Test
 X²=35.41666667
 P=3.8142016E-7
 df=4
```

5. The 3×2 contingency table gives the observed frequencies f_{ij}. There is no way of calculating the expected frequencies using theoretical probabilities, so we must follow the method of *Worked Example* 2. We calculate $R_1 = 2004$, $R_2 = 205$, and $R_3 = 28$. Also, $C_1 = 1170$ and $C_2 = 1067$. We use the formula $e_{ij} = R_iC_j/2237$ to calculate the values expected frequencies. (cont'd)

Exploration 19

5. (cont'd) We then compute

$$\chi_2^{\,2} = \sum_{i=1}^{3}\sum_{j=1}^{2} \frac{(f_{ij} - e_{ij})^2}{e_{ij}}$$

$$= \frac{(1070-1048)^2}{1048} + \frac{(934-956)^2}{956} + \frac{(92-107)^2}{107} + \frac{(113-98)^2}{98} + \frac{(8-15)^2}{15} + \frac{(20-13)^2}{13}$$

$= 11.806\ldots$

The display shows that χ^2_2 takes a value of 11.806 or greater with a probability of about 0.00273.

```
X²cdf(11.806,E99
,2)
          .0027312388
```

Therefore, we can reject the null hypothesis at the 1% level of significance. That is, there is a higher proportion of right-handedness among women than among men.

Alternatively, we can use the χ^2 test on the TI-83 (see *Investigation* 9) and define the matrix [A] as shown in the display.

```
MATRIX[A]  3 ×2
[ 1070   934   ]
[ 92     113   ]
[ 8      20    ]
```

The display verifies the result obtained above.

```
X²-Test
 X²=11.80613467
 P=.0027310549
 df=2
```

CHALLENGE

a) X_1 is a random variable with a binomial distribution having mean np (where p is the probability of that outcome on each toss) and standard deviation $\sqrt{np(1-p)}$. As the number of tosses becomes large, this distribution is closely approximated by a normal distribution with $\mu = np$ and $\sigma = \sqrt{np(1-p)}$.

b) Therefore the random variable $\frac{X_1 - np}{\sqrt{np(1-p)}}$ is normally distributed with $\mu = 0$ and $\sigma = 1$.

c) Hint: From part b) and the theorem on page 64, $\frac{(X_1 - np)^2}{np(1-p)}$ has a chi-square distribution with 1 degree of freedom. Furthermore,

$$\frac{1}{p(1-p)} = \frac{1}{p} + \frac{1}{1-p}, \text{ so } \frac{(X_1 - np)^2}{np(1-p)} = \frac{(X_1 - np)^2}{np} + \frac{(X_1 - np)^2}{n(1-p)}.$$

Exploration 20

1. No. The diets may have promoted substantially different weight losses. However, the sample data indicates that while the means are different, their differences are within the range of chance variation and therefore, the null hypothesis cannot be rejected.

3. b) $\sum_{j=1}^{n} (x_{ij} - \bar{x}_i)^2 = (n-1)s_i^2$

4. b) $\sum_{j=1}^{16} (x_{2j} - \bar{x}_2)^2 = 15(7.501111)^2 = 844$

5. $SS_w = 4483.7$ $\quad SS_b = 3518.6$

$F = 3(n-1)SS_b / 2SS_w$

$= 10.59417\ldots$

```
Fcdf(10.59417,E9
9,2,27)
       4.015080679E-4
```

Since $P \approx 4 \times 10^{-4}$, we can reject the null hypothesis at the 1% level of significance.

Copyright © 1998 by Brendan Kelly Publishing Inc.